Lectures on Discrete Time Filtering

Signal Processing and Digital Filtering

Synthetic Aperture Radar
J.P. Fitch

Multiplicative Complexity, Convolution and the DFT
M.T. Heideman

Array Signal Processing
S.U. Pillai

Maximum Likelihood Deconvolution
J.M. Mendel

Algorithms for Discrete Fourier Transform and Convolution
R. Tolimieri, M. An, and C. Lu

Algebraic Methods for Signal Processing and Communications Coding
R.E. Blahut

Electromagnetic Devices for Motion Control and Signal Processing
Y.M. Pulyer

Mathematics of Multidimensional Fourier Transform Algorithms
R. Tolimieri, M. An, and C. Lu

Lectures on Discrete Time Filtering
R.S. Bucy

R.S. Bucy

with the assistance of
B.G. Williams

Lectures on Discrete Time Filtering

C.S. Burrus
Consulting Editor

With 15 Illustrations

Springer-Verlag
New York Berlin Heidelberg London Paris
Tokyo Hong Kong Barcelona Budapest

R.S. Bucy
Department of Aerospace Engineering
University of Southern California
Los Angeles, CA 90089 USA

B.G. Williams
Jet Propulsion Laboratory
California Institute of Technology
Pasadena, CA 91109 USA

Consulting Editor
Signal Processing and Digital Filtering

C.S. Burrus
Professor and Chairman
Department of Electrical
 and Computer Engineering
Rice University
Houston, TX 77251-1892
USA

Library of Congress Cataloging-in-Publication Data
Bucy, Richard S., 1935–
  Lectures on discrete time filtering / R.S. Bucy.
    p.   cm. — (Signal processing and digital filtering)
  Includes bibliographical references.
  ISBN 0-387-94198-3. — ISBN 3-540-94198-3 (Berlin)
  1. Electric filters—Mathematical models.   2. Signal processing—
Mathematics.   3. Discrete-time systems—Mathematical models.
  I. Title.   II. Series.
TK7872.F5B85 1994
003'.83'01154—dc20   93-43310

Printed on acid-free paper.

© 1994 Springer-Verlag New York, Inc.
All rights reserved. This work may not be translated or copied in whole or in part without the written permission of the publisher (Springer-Verlag New York, Inc., 175 Fifth Avenue, New York, NY 10010, USA), except for brief excerpts in connection with reviews or scholarly analysis. Use in connection with any form of information storage and retrieval, electronic adaptation, computer software, or by similar or dissimilar methodology now known or hereafter developed is forbidden.
The use of general descriptive names, trade names, trademarks, etc., in this publication, even if the former are not especially identified, is not to be taken as a sign that such names, as understood by the Trade Marks and Merchandise Marks Act, may accordingly be used freely by anyone.

Production managed by Hal Henglein; manufacturing supervised by Jacqui Ashri.
Photocomposed copy prepared from the author's LaTeX files.
Printed and bound by Edwards Brothers, Inc., Ann Arbor, MI.
Printed in the United States of America.

9 8 7 6 5 4 3 2 1

ISBN 0-387-94198-3 Springer-Verlag New York Berlin Heidelberg
ISBN 3-540-94198-3 Springer-Verlag Berlin Heidelberg New York

Preface

The theory of linear discrete time filtering started with a paper by Kolmogorov in 1941. He addressed the problem for stationary random sequences and introduced the idea of the innovations process, which is a useful tool for the more general problems considered here. The reader may object and note that Gauss discovered least squares much earlier; however, I want to distinguish between the problem of parameter estimation, the Gauss problem, and that of Kolmogorov estimation of a process. This separation is of more than academic interest as the least squares problem leads to the normal equations, which are numerically ill conditioned, while the process estimation problem in the linear case with appropriate assumptions leads to uniformly asymptotically stable equations for the estimator and the gain. The conditions relate to controlability and observability and will be detailed in this volume.

In the present volume, we present a series of lectures on linear and nonlinear sequential filtering theory. The theory is due to Kalman for the linear colored observation noise problem; in the case of white observation noise it is the analog of the continuous-time Kalman–Bucy theory. The discrete time filtering theory requires only modest mathematical tools in counterpoint to the continuous time theory and is aimed at a senior-level undergraduate course. The present book, organized by lectures, is actually based on a course that meets once a week for three hours, with each meeting constituting a lecture. The course has been given at the University of Southern California, Los Angeles; the University of Nice; and Cheng Kung University, Tainan, Taiwan.

The theory presented here, especially in the linear case, has had numerous applications in many areas; to name a few: economics—estimation of the money supply; geophysics—sonar processing; numerical analysis—integration packages; guidance—space vehicles, Apollo; fire control—F14, F15, and helicopter systems; location—AWACS, GPS; space—space station.

A number of these and other applications are referenced in R.S. Bucy and P.D. Joseph, "Filtering for Stochastic Processes with Applications to Guidance," 2d edition, Chelsea, New York, 1987. These applications were possible because of the wide applicability of Markov models and because the linear theory yields recursive equations for the estimator and the gains.

The nonlinear theory is treated along with some novel synthesis methods for this computationally demanding problem. The Burg technique, the Schmidt multiple-signal classification, MUSIC, and a detailed analysis of the asymptotic theory of the matrix Riccati equation are also covered. Knowledge of the behavior of the Riccati equation is indispensable for practical applications. Finally, only a small number of problems are given in the text as we feel that an important part of the learning process for the student is to construct examples that illustrate and illuminate the text.

The model for the form of this book is the *Lectures on Dynamics* by Jacobi. We will feel that we have been successful if we can approach the elegance of that book.

Los Angeles, California   R.S. Bucy
December, 1993

Contents

Preface			v
Introduction			xiii

1 Review — 1
1   Review of Concepts in Probability 1
     1.1   Gaussian Random Variables 5

2 Random Noise Generation — 13
1   Random Noise Generation 13
2   Cholesky Decomposition 16
3   Uses of the Pseudo Inverse 17
4   Signal Models 19
5   Sensor Model 21

3 Historical Background — 23
1   Background Material 23
2   Historical Developments for Filtering 24
     2.1   Concept of Innovations 24
     2.2   Wiener-Hopf Equation 26
3   Development of Innovations 30
4   Sequential Filter Development 31

4 Sequential Filtering Theory — 37
1   Summary of the Sequential Filter 37
2   The Scalar Autonomous Riccati Equation 39
3   Linearizing the Riccati Equation 43

		3.1	Symplectic Matrices	44
		3.2	Stability of the Filter	45

5 Burg Technique — 47
1. Background Material — 47

6 Signal Processing — 55
1. The Burg Technique — 55
2. Signal Processing — 58
3. Burg Revisited (Rouché's Theorem) — 61
   - 3.1 Burg's Inverse Iteration — 63

7 Classical Approach — 71
1. Classical Steady-State Filtering — 71

8 A Priori Bounds — 81
1. A Priori Bounds for the Riccati Equation — 81
2. Information and Filtering — 89
3. Nonlinear Systems — 91

9 Asymptotic Theory — 93
1. Applications of the Theory of Filtering — 93
2. Asymptotic Theory of the Riccati Equation — 96
3. Steady-State Solution to Riccati — 100

10 Advanced Topics — 107
1. Invariant Directions — 109
2. Nonlinear Filtering — 110

11 Applications — 117
1. Historical Applications — 117
   - 1.1 Problem 1. Cubic Sensor Problem ($d=1$) — 118
   - 1.2 Numerical Realization — 121
   - 1.3 Problem 2. Passive Receiver — 123

12 Phase Tracking — 127
   1. The Phase Lock Loop 127
   2. Phase Demodulation 130

13 Device Synthesis — 133
   1. Device Synthesis for Nonlinear Filtering 133
      - 1.1 Hybrid Computing in Nonlinear Filtering 133
      - 1.2 Optical Techniques for Nonlinear Filter Convolution 135
      - 1.3 Acoustic Techniques for Nonlinear Filter Convolution 138
      - 1.4 Digital Developments in Nonlinear Filtering 138
   2. Radar Filtering Application 140
      - 2.1 Rao-Cramer Bound 142

14 Random Fields — 145

Bibliography — 151

List of Figures

4.1	System diagram for the linear filter.	38
4.2	Phase diagram for scalar Riccati equation.	40
7.1	Block diagram of the O–U process.	74
10.1	Block diagram of the process in the new basis.	109
10.2	System diagram for the nonlinear filter.	115
11.1	Two-dimensional passive receiver geometry.	124
13.1	Block diagram of the nonlinear filter showing segmentation A for convolution and B for data processing.	134
13.2	Schematic of convolution as a dot product.	134
13.3	Generation of new Ps with hybrid computing hardware.	135
13.4	Schematic for incoherent optical convolution.	136
13.5	Schematic of a single scalar convolution for Mallinckrodt's TV idea.	137
13.6	Schematic of the surface acoustic wave convolution device.	138
13.7	Phased array radar filtering problem.	140
13.8	Ordered eigenvalues of R.	142
13.9	Estimation of radar angles using the Schmidt statistic.	142

Introduction

In these lectures our purpose is to expose not only the theory of discrete time filtering, but also to provide a firm basis for understanding the theory in terms of rigorous discussion of random variables, random processes, conditional probability, and differential equations. Our hope is to provide insight and understanding into the methods and properties of the solutions, rather than the more usual cookbook listing of the equations. We believe that in applications a firm foundation in the theory is indispensable. Further, for those who wish to attack new problems a deep understanding of theory is clearly required.

The first lecture reviews probability theory and the measure theory necessary to precise understanding of the concepts of random variable, expectation and various limit theorems. Properties of linear Gaussian random variables are delineated.

The second lecture provides a "white noise" generator that is machine-independent. Since a large part of the book concerns random process models that consist of the output of linear systems driven by white noise, this discussion gives the reader the tools to realize a large class of random processes on a digital computer. The Cholesky decomposition is derived, which provides a tool to pass from a covariance description of a process to a state variable one; i.e., the output of a linear system driven by white noise.

The third lecture derives the linear discrete time optimal filter for a finite-dimensional Markov Gaussian signal process partial observed in the presence of additive white noise. We carefully distinguish the special prediction problem (no observation noise), from the more general filtering problem.

The fourth lecture deals with the block diagram structure of the optimal filter and the Riccati equation, the equation that not only determines the "gains" in the optimal filter but also determines the error covariance of the optimal filter. In particular, the one-dimensional Riccati equation theory is developed to motivate the reader to understand the theory of the matrix equation, which is presented later.

The fifth and sixth lectures are devoted to the Burg solution to the special prediction problem. We show how this problem can be related to a special case of the filtering problem. The Burg technique, a process that converts data into a spectrum, is widely used and can be understood as an approximation to a special case of the optimal filter. Programs in BASIC that realize various Burg spectrum estimators are included.

The seventh lecture compares the Wiener spectral factorization to the methods of sequential filtering. We relate the steady-state solution to the Riccati equation to spectral factors of the spectrum of the observation process.

Lecture eight derives a priori bounds for the Riccati equation, which are the key to understanding the asymptotic behavior, stability, and dynamics of the optimal filter and the Riccati equation. Requisite regularity conditions for nice behavior of the filter, uniform complete controlability, and observability are discussed. Connections with information theory are delineated.

Lecture nine discusses the asymptotic behavior of the Riccati equation. This theory, although often quoted, is never proved in detail, and here we fill in the gaps in the literature. This theory is indispensable for filter realization debugging as it paints a qualitative picture of the behavior of solutions of the Riccati equation. The important Bass-Roth theorem, the accepted method for finding steady-state solutions to the autonomous Riccati equation, is discussed.

Lecture ten considers the dynamic dimension of the Riccati equation for the "colored" noise problem. This is related to the instability of the normal equations. The derivation of the equations for the important nonlinear filtering problem is given. The nonlinear filter block diagram is also given.

Lecture eleven considers the synthesis of the optimal nonlinear filter

Introduction

for two historically important problems, the cubic sensor and the passive receiver. Both of these problems were analyzed early by Monte Carlo analysis as they both had low state dimension. The equations of the nonlinear filter have been synthesized by using the white noise generator on modern high-speed digital computers, and some of these realizations are reviewed further in lecture thirteen.

The problem of determining the optimal nonlinear filter for phase demodulation is considered in lecture twelve. This problem is important in space communications and submarine communications. The performance of the optimal filter, obtained by Monte Carlo simulation, is compared to the "cyclic" performance of the phase lock loop, which is really an extended Kalman-Bucy filter.

Lecture thirteen considers how to overcome the need for massive computation to realize the optimal nonlinear filter. Parallel and analog computation methods are investigated and the results are reported for surface acoustic wave devices, optical convolutions, and various parallel digital architecture. An example of a high state dimensional problem of adaptive radar arrays, for which the optimal nonlinear filter realization is beyond current capabilities, is given along with a suboptimal technique.

Another area where high state dimension occurs is picture processing. It is considered in lecture fourteen where the natural model is that of random fields.

This book has developed from lectures given for a one-semester course in advanced topics of the aerospace engineering department at the University of Southern California. The assistance provided by two students of that course, P.W. Chodas and T.P. McElrath, to compile these lecture notes in their current form is hereby gratefully acknowledged.

Lecture 1

Review

1 Review of Concepts in Probability

Consider a space made up of points $\Omega = \{\omega_\alpha\}$ where each ω_α is an elementary event. The event $A \subseteq \Omega$ is a collection of ω_αs. Ω is called the sure event, or certain event, and $\emptyset$ is the empty set.

The usual logical set operations apply:

Union: $A \cup B = \{\omega \,|\, \omega \in A, \text{ or } \omega \in B, \text{ or both}\}$

Intersection: $A \cap B = \{\omega \,|\, \omega \in A \text{ and } \omega \in B\}$

Complement: $A^c = \bar{A} = \{\omega \,|\, \omega \notin A\}$

$$\limsup A_n = \bigcap_{n=1}^{\infty} \bigcup_{k \geq n} A_k = \{\omega \,|\, \omega \in A_k \text{ for infinitely many } ks\}$$

$$\liminf A_n = \bigcup_{n=1}^{\infty} \bigcap_{k \geq n} A_k = \{\omega \,|\, \omega \in A_k \text{ eventually}\}.$$

The sequence A_n has a limit iff $\limsup A_n = \liminf A_n$. These operations can be expressed in terms of Venn diagrams. Given a set of events it is possible to make more events using these operations.

Now define $\mathcal{A}$ as a collection of events with closure properties or a sigma field. It is closed under countable set operations, and $\Omega \in \mathcal{A}$. For this sigma field, define a map P such that

$$P : \mathcal{A} \mapsto [0, 1].$$

Some properties of P are:

1. $P(\emptyset) = 0$, and

2. $P(\bigcup A_i) = \sum P(A_i)$, where $A_i \cap A_j = \emptyset$ for $i \neq j$; N.B. the index i may take countable values.

Example 1.1 *These three entities define a probability space $(\Omega, \mathcal{A}, P)$. Suppose a coin is tossed three times. Then there are eight possibilities for different combinations of heads and tails. $\Omega = (\omega_1, \omega_2, ..., \omega_8)$ where each ω_i corresponds to one of the elementary events, i.e., outcomes of the coin-tossing experiment. $\mathcal{A}$ is the set of all subsets of Ω, of which there are 2^α, where α is the number of elementary events. Suppose that the probability of each ω_i is $1/8$, find the probability of two heads.*

Definition 1.1 *A map $X : \Omega \mapsto R$ such that $X^{-1}(a,b) \in \mathcal{A}$ for all $a,b \in R$ is a random variable or measurable function.*

$$X^{-1}(a,b) = \Big(\omega \mid a < X(\omega) < b\Big).$$

It is useful to generalize the notion of random variable as follows; consider the equivalence relation:

$$X \equiv Y$$

iff $(X - Y)$ differs from zero only on a set of probability zero; i.e., a.e.p. (*almost everywhere in probability*). A random variable will be a representative of the equivalence class. Any function is a random variable when the cardinality of Ω is finite and $\mathcal{A}$ is the collection of all subsets of Ω. Notice that by definition of random variable, it is a measurable real valued function with respect to $\mathcal{A}$, so that pointwise limits of random variables are random variables.

Two types of convergence for sequences, $X_n \to X$, are useful:

1. Almost sure convergence iff $P(\limsup X_n > \liminf X_n) = 0$.

2. Convergence in probability iff $P(|X_n - X| > \epsilon) \to 0$ for arbitrary $\epsilon > 0$.

Almost sure convergence is abbreviated a.s. and is convergence almost everywhere in measure theory. In fact, it is convenient to deal with the equivalence class of random variables that differ on a null set or are a.s. equal. Convergence in probability is called convergence in measure in measure theory. If $X_n \to X$ in measure then there exists a subsequence that converges a.e., i.e., almost everywhere. Convergence in probability is abbreviated a.e.p.

Example 1.2 *If $\mathcal{A} = \{\Omega, \emptyset\}$, then random variables are only the functions $X(\omega) = constant$ a.e.p.*

Large $\mathcal{A}$ allows many random variables. For $\Omega = (-\infty, \infty)$ the smallest $\mathcal{A}$ containing

$$A_{r,s} = \{\omega \mid r < \omega < s\}$$

for all $r, s \in$ rationals is called the Borel field, and the measure that assigns to an interval its length is the Lebesgue measure.

Often one is interested in the ensemble average. This is accomplished by integration with respect to Ω. This ensemble average is called the expected value, denoted $E\,g(X)$, and $g : R \mapsto R$ (g is a Borel measurable map from R to R, while X is a random variable).

Now, define the expected value operator, E, as follows. If

$$X(\omega) = \sum_{i=1}^{n} a_i I_{A_i}(\omega),$$

where $A_i \in \mathcal{A}$, and $\bigcup A_i = \Omega$ with $A_i \cap A_j = \emptyset$ for $i \neq j$ and with

$$I_{A_i}(\omega) = \begin{cases} 1, & \text{if } \omega \in A_i \\ 0, & \text{otherwise} \end{cases},$$

which is the indicator function of the set A_i, then

$$EX(\omega) = \sum_{i=1}^{n} a_i P(A_i).$$

The following are properties of E:

$$X \geq 0 \Rightarrow EX \geq 0,$$

$$E(\alpha X + \beta Y) = \alpha EX + \beta EY$$

for X and Y measurable. Approximations to measurable and positive X are given by

$$X_n = \sum_{k=0}^{n2^n-1} \frac{k}{2^n} I_{[k/2^n < x(\omega) \leq (k+1)/2^n]} + n I_{[x(\omega) > n]},$$

and $X > X_n$, and X_n is a simple function monotone, which converges up to X.

The expectation can now be defined as $\lim EX_n = EX = \int_\Omega X \, dP$. The integral is called the Lebesgue integral, which works by stepping along the dependent axis (as opposed to the Riemann integral, which steps along the independent axis). The integral or expectation can be extended to all random variables for which either $E(X^+)$ or $E(X^-)$ is finite, where X^+ and X^- are non-negative and $X = X^+ - X^-$. For complete details on the development of the integral, see [1]. In particular, the following theorems are useful. Let X_n be non-negative random variables, then we have the following.

Theorem 1.1 (Monotone Convergence) $X_n \uparrow X$ *implies* $E(X_n) \uparrow E(X)$.

Theorem 1.2 (Dominated Convergence) $|X_n| \leq B$ *and* $X_n \to X$ *with* $E(B) < \infty$ *implies* $E(X_n) \to E(X)$.

P maps sets to R, which is hard to deal with because of the complexity of the domain. Introduce an equivalent point function when a random

1. Review of Concepts in Probability

variable X is studied as

$$P(x < a) = F^x(a),$$

which is called the distribution function of x. Some properties of the distribution function are

$$F^x(-\infty) = 0,$$

$$a \leq b \Rightarrow F^x(a) \leq F^x(b),$$

and F^x is continuous from the left or right as a consequence of the sigma additivity of the probability measure, depending on whether $<$ or $\leq$ is used in the definition of F^x. An arbitrary distribution function can be decomposed into three components, a set of point masses, an absolutely continuous part (the indefinite integral of a density), and a singular part; see [1] for details.

1.1 Gaussian Random Variables

In much of what follows, we will be interested in the properties of the Gaussian density. For a Gaussian distribution,

$$P(x \leq y) = \int_{-\infty}^{y} \frac{1}{(2\pi\sigma^2)^{1/2}} e^{-(x-\mu)^2/2\sigma^2} dx$$

$$\mathrm{E}x = \mu \quad \text{and} \quad \mathrm{E}(x-\mu)^2 = \sigma^2.$$

The integrand defining the Gaussian distribution is related to the Green's function of the heat equation and Brownian motion. We will denote the assumption that x has a Gaussian distribution with mean μ and variance σ^2 as $x \in N(\mu, \sigma^2)$. The following two lemmas describe the Gaussian distribution:

Lemma 1.1

$$\int_{-\infty}^{\infty} \frac{1}{(2\pi\sigma^2)^{1/2}} e^{-(x-\mu)^2/2\sigma^2} dx = 1.$$

This can be proved by squaring the integral and changing to polar coordinates.

Lemma 1.2

$$\int_{-\infty}^{\infty} \frac{x}{(2\pi\sigma^2)^{1/2}} e^{-(x-\mu)^2/2\sigma^2} dx$$

$$= \int_{-\infty}^{\infty} \frac{x-\mu}{(2\pi\sigma^2)^{1/2}} e^{-(x-\mu)^2/2\sigma^2} dx$$

$$+ \mu \int_{-\infty}^{\infty} \frac{1}{(2\pi\sigma^2)^{1/2}} e^{-(x-\mu)^2/2\sigma^2} dx = \mu$$

because the first term is odd and vanishes, and the integral of the second term is 1 by Lemma 1.1. The characteristic function is

$$\phi(i\omega) = Ee^{i\omega x}$$

$$= \int_{-\infty}^{\infty} \frac{e^{i\omega x} e^{-(x-\mu)^2/2\sigma^2}}{(2\pi\sigma^2)^{1/2}} dx$$

$$= \int_{-\infty}^{\infty} \frac{1}{(2\pi\sigma^2)^{1/2}} e^{\left(\frac{-x^2}{2\sigma^2} + \frac{\mu x}{\sigma^2} + i\omega x - \frac{\mu^2}{2\sigma^2}\right)} dx.$$

By completing the square in the exponent

$$\int_{-\infty}^{\infty} \frac{1}{(2\pi\sigma^2)^{1/2}} e^{-(x-\mu-i\omega\sigma^2)^2/2\sigma^2} e^{\frac{-\mu^2}{2\sigma^2}} e^{\frac{(\mu-i\omega\sigma^2)^2}{2\sigma^2}} dx = e^{i\omega\mu - \frac{\omega^2\sigma^2}{2}}.$$

This is the characteristic function from which all of the moments can be obtained by differentiating. From this the n-dimensional Gaussian characteristic function can be obtained by noting that the inner product $l'\mathbf{x}$ is a scalar Gaussian random variable with an arbitrary constant vector l, so that the characteristic function for a Gaussian random vector is determined from that of $l'\mathbf{x}$. The n-dimensional Gaussian density function is

$$p(\mathbf{x}) = \frac{1}{(2\pi)^{n/2} \det \Sigma^{1/2}} e^{-\frac{1}{2}\|\mathbf{x}-\mu\|^2_{\Sigma^{-1}}} d\mathbf{x},$$

1. Review of Concepts in Probability

where $\Sigma = E(\mathbf{x}-\mu)(\mathbf{x}-\mu)'$, also called the covariance, and where $\|\mathbf{x} - \mu\|^2_{\Sigma^{-1}} = (\mathbf{x} - \mu)'\Sigma^{-1}(x - \mu)$ is called the "norm squared with respect to sigma inverse."

Now, define the conditional probability for $(\Sigma, \mathcal{A}, P)$: If $\mathcal{B} \subseteq \mathcal{A}$ then $P(A|\mathcal{B}) = E\, I_A \mid \mathcal{B}$. $P(A \mid \mathcal{B})$ is the probability of A given $\mathcal{B}$. $EX|\mathcal{B}$ to be $\mathcal{B}$ measurable.

$$\int_B X\, dP = \int_B E(X \mid \mathcal{B}) dP_{\mathcal{B}}, \quad \text{for } \forall\ B \text{ in } \mathcal{B}.$$

If X is integrable, there is a $\mathcal{B}$ measurable solution to this equation; this is the content of the Radon Nikodym theorem (see [1]).

Filtering is determination of the conditional distribution of the signal given the smallest sigma field for which the observations are measurable. Further realization of the optimal filter is just the construction of a black box that accepts as inputs observations and outputs the conditional density of the signal at time t given the observations up to t.

Suppose x and y are zero mean Gaussian random variables with density $p(\bullet, \bullet)$. Then

$$p(x,y) = \frac{1}{2\pi(\sigma_{11}^2 \sigma_{22}^2 - \sigma_{12}^4)^{1/2}} e^{-\frac{1}{2D}(\sigma_{22}^2 x^2 - 2\sigma_{12}^2 xy + \sigma_{11}^2 y^2)},$$

where $\sigma_{11}^2 = Ex^2, \sigma_{22}^2 = Ey^2$, and $\sigma_{12}^2 = Exy$. This follows from

$$\Sigma = \begin{pmatrix} \sigma_{11}^2 & \sigma_{12}^2 \\ \sigma_{12}^2 & \sigma_{22}^2 \end{pmatrix}$$

and

$$\Sigma^{-1} = \frac{1}{D} \begin{pmatrix} \sigma_{22}^2 & -\sigma_{12}^2 \\ -\sigma_{12}^2 & \sigma_{11}^2 \end{pmatrix},$$

where

$$D = (\sigma_{11}^2 \sigma_{22}^2 - \sigma_{12}^4).$$

The conditional density of x given y is

$$p(x|y) = \frac{p(x,y)}{p(y)}$$

$$= \frac{(2\pi\sigma_{22}^2)^{\frac{1}{2}} e^{-\frac{1}{2D}(\sigma_{22}^2 x^2 - 2\sigma_{12}^2 xy + \sigma_{11}^2 y^2)}}{2\pi D^{1/2} \, e^{\frac{-D}{2\sigma_{22}^2 Dy^2}}}$$

$$= \frac{1}{(2\pi)^{1/2}(\frac{D}{\sigma_{22}^2})^{1/2}} e^{-\frac{\sigma_{22}^2}{2D}(x^2 - 2\sigma_{12}^2 xy + \frac{1}{\sigma_{22}^2}(\sigma_{11}^2 - \frac{D}{\sigma_{22}^2})y^2)}$$

$$= \frac{1}{(2\pi)^{1/2}(\frac{D}{\sigma_{22}^2})^{1/2}} e^{-\frac{\sigma_{22}^2}{2D}(x - \frac{\sigma_{12}^2}{\sigma_{22}^2} y)^2}.$$

This is a normal distribution with

$$N\left(\frac{\sigma_{12}^2}{\sigma_{22}^2} y, \sigma_{11}^2 - \frac{\sigma_{12}^2 \sigma_{12}^2}{\sigma_{22}^2}\right).$$

So X, Y Gaussian implies $p(x \mid y)$ Gaussian. In addition, μ is linear in y, and σ^2 is independent of y. Describing it in a slightly different way,

$$\mathrm{E}(x|y) = \mathrm{E}xy(\mathrm{E}yy)^{-1}y$$

and

$$\mathrm{E}(x - \mathrm{E}x|y)^2 = \mathrm{E}xx - \mathrm{E}xy(\mathrm{E}yy)^{-1}\mathrm{E}yx.$$

To generalize this result for an n-dimensional distribution, we need to use the Schur determinant relations [2]. Let

$$\Upsilon = \begin{pmatrix} A & B \\ C & D \end{pmatrix},$$

where A is an $s \times s$ matrix and D is an $(n-s) \times (n-s)$ matrix. The determinant of Υ can be found in terms of $A, B, C,$ and D in the following way when A and D are invertible:

$$|\Upsilon| = \left|\begin{pmatrix} A & B \\ C & D \end{pmatrix}\begin{pmatrix} I & -A^{-1}B \\ 0 & I \end{pmatrix}\right|$$

$$= \left|\begin{pmatrix} A & 0 \\ C & D-CA^{-1}B \end{pmatrix}\right| = |A||D-CA^{-1}B|,$$

or by symmetry

$$|\Upsilon| = |D||A - BD^{-1}C|.$$

From this one can easily arrive at the relation

$$\Upsilon^{-1} = \begin{pmatrix} (A-BD^{-1}C)^{-1} & -A^{-1}B(D-CA^{-1}B)^{-1} \\ -D^{-1}C(A-BD^{-1}C)^{-1} & (D-CA^{-1}B)^{-1} \end{pmatrix},$$

which can be verified by direct premultiplication by Υ. By noting that in any group a right inverse is a left inverse, the following useful relations can be derived:

$$(A - BD^{-1}C)^{-1} = A^{-1} + A^{-1}B(D - CA^{-1}B)^{-1}CA^{-1}$$

$$(D - CA^{-1}B)^{-1} = D^{-1} + D^{-1}C(A - BD^{-1}C)^{-1}BD^{-1}$$

$$(A - BD^{-1}C)^{-1}BD^{-1} = A^{-1}B(D - CA^{-1}B)^{-1}$$

$$(D - CA^{-1}B)^{-1}CA^{-1} = D^{-1}C(A - BD^{-1}C)^{-1}.$$

Now we find going to the conditional density for the vector case: for $\mathbf{x}$, $\mathbf{y}$ that are $N(0, \boldsymbol{\Sigma})$,

$$p(\mathbf{z}) = p(\mathbf{x}, \mathbf{y}) = \frac{1}{(2\pi)^{n/2}|\boldsymbol{\Sigma}|^{1/2}} \exp\left(-\frac{1}{2}\|\mathbf{z}\|_{\boldsymbol{\Sigma}^{-1}}^2\right),$$

where

$$\Sigma = \begin{pmatrix} \Sigma_{11} & \Sigma_{12} \\ \Sigma_{21} & \Sigma_{22} \end{pmatrix}$$

and $\Sigma_{ij} = E z_i z_j'$, where $z_1 = x$ and $z_2 = y$.

$$\Sigma^{-1} = \begin{pmatrix} (\Sigma_{11} - \Sigma_{12}\Sigma_{22}^{-1}\Sigma_{21})^{-1} & -\Sigma_{11}^{-1}\Sigma_{12}(\Sigma_{22} - \Sigma_{21}\Sigma_{11}^{-1}\Sigma_{12})^{-1} \\ -\Sigma_{22}^{-1}\Sigma_{21}(\Sigma_{11} - \Sigma_{12}\Sigma_{22}^{-1}\Sigma_{21})^{-1} & (\Sigma_{22} - \Sigma_{21}\Sigma_{11}^{-1}\Sigma_{12})^{-1} \end{pmatrix}.$$

Let $L = (\Sigma_{11} - \Sigma_{12}\Sigma_{22}^{-1}\Sigma_{21})^{-1}$ and $M = (\Sigma_{22} - \Sigma_{21}\Sigma_{11}^{-1}\Sigma_{12})^{-1}$. Then

$$p(\mathbf{x} \mid \mathbf{y}) = \frac{\frac{1}{(2\pi)^{n/2}|\Sigma|^{1/2}} e^{-\frac{1}{2}(\mathbf{x}L\mathbf{x}' - 2\mathbf{y}'\Sigma_{22}^{-1}\Sigma_{21}L\mathbf{x} + \mathbf{y}'M\mathbf{y})}}{\frac{1}{(2\pi)^{n_y/2}|\Sigma_{22}|^{1/2}} e^{-\frac{1}{2}\mathbf{y}'\Sigma_{22}^{-1}\mathbf{y}}}.$$

Using the Schur relations,

$$p(\mathbf{x}|\mathbf{y}) \frac{1}{(2\pi)^{\frac{n_x}{2}} \left[\Sigma_{11} - \Sigma_{12}\Sigma_{22}^{-1}\Sigma_{21}\right]^{1/2}} e^{-\frac{1}{2}\left(\mathbf{x}L\mathbf{x}' - 2\mathbf{y}'\Sigma_{22}^{-1}\Sigma_{21}L\mathbf{x} + \mathbf{y}'M\mathbf{y} - \mathbf{y}'\Sigma_{22}^{-1}\mathbf{y}\right)}.$$

Looking at the last two terms of the exponent,

$$\mathbf{y}'M\mathbf{y} - \mathbf{y}'\Sigma_{22}^{-1}\mathbf{y} = \mathbf{y}'\Sigma_{22}^{-1}\Sigma_{21}L\Sigma_{21}\Sigma_{22}^{-1}\mathbf{y},$$

also by the Schur relations. This is now a perfect square, so

$$p(\mathbf{x}|\mathbf{y}) = \frac{1}{(2\pi)^{\frac{n_x}{2}}|\Sigma_{11} - \Sigma_{12}\Sigma_{22}^{-1}\Sigma_{21}|^{1/2}} e^{-\frac{1}{2}\|\mathbf{x} - \Sigma_{12}\Sigma_{22}^{-1}\mathbf{y}\|_L^2}.$$

Theorem 1.3 *If* $\mathbf{y}$ *and* $\mathbf{y}$ *are Gaussian, zero-mean n and m vectors such that*

$$E\begin{pmatrix} \mathbf{x} \\ \mathbf{y} \end{pmatrix}(\mathbf{x}' \ \mathbf{y}') = \begin{pmatrix} \Sigma_{11} & \Sigma_{12} \\ \Sigma_{21} & \Sigma_{22} \end{pmatrix},$$

then

$$P^{\mathbf{x}|\mathbf{y}}(\mathbf{a}) = \int_{-\infty}^{a_1} \cdots \int_{-\infty}^{a_{n_x}} \frac{1}{(2\pi)^{n_x/2}|\Sigma_{11} - \Sigma_{12}\Sigma_{22}^{-1}\Sigma_{21}|^{1/2}} e^{-\frac{1}{2}\|\mathbf{x} - \Sigma_{12}\Sigma_{22}^{-1}\mathbf{y}\|_L^2} d\mathbf{x},$$

where $L^{-1} = (\Sigma_{11} - \Sigma_{12}\Sigma_{22}^{-1}\Sigma_{21})$.

From the $-2\mathbf{y}'\Sigma_{22}^{-1}\Sigma_{21}L\mathbf{x}$ term of the exponent one sees that if $\mathbf{x}, \mathbf{y}$ are Gaussian and uncorrelated ($\Sigma_{21} = 0$), then $\mathbf{x}, \mathbf{y}$ are independent ($P(\mathbf{x}, \mathbf{y}) = P(\mathbf{x})P(\mathbf{y})$).

One more note on the matrix inverse built from the Schur relations: it was assumed that A and D were invertible. But, AA' is positive definite, and $A^{-1} = A'(AA')^{-1}$, so it can always be made to work.

Definition 1.2 *A random process is an indexed collection of random variables.*

Remark: The characteristic function for Gaussian vectors shows that if $\mathbf{x}$ and $\mathbf{y}$ are Gaussian and independent, $\mathbf{z} = a\mathbf{x} + b\mathbf{y}$ is Gaussian. Using the Gram-Schmidt procedure it is easy to see that Gaussian random variables are closed under linear combination.

Problem 1.1 *If z is a complex random variable with Gaussian real and imaginary parts, find the distributions of the phase and amplitude of z.*

Lecture 2

Random Noise Generation

1 Random Noise Generation

If P is a covariance then P is positive semidefinite and $P = T'\Lambda T$ with T orthogonal and $\lambda_i \geq 0$. The n-dimensional Gaussian integral is defined and the normalization satisfies

$$(2\pi)^{n/2}|P|^{1/2} = \int e^{-1/2\|x-\mu\|^2_{P^{-1}}} dx$$
$$= (2\pi)^{n/2}(\lambda_1 \cdots \lambda_n)^{1/2}.$$

In order to construct uniform random variables, consider integer sequences such as $l_n \in [0, 2^m]$. Since the values of random variables must remain between 0 and 1, $l_n/2^m$ is a candidate for an approximate uniformly distributed pseudo random variable. Of course, the mapping $l : I \to [0, 2^m]$ must be chosen with care to provide a good approximation.

Von Neumann suggested the pseudo random sequence

$$l_{n+1} = l_n^2 \pmod{2^m}$$

for the construction of the uniform random variables. IBM synthesized Gaussian random variables from independent uniformly distributed random variables on [0,1], X_i, via the model:

$$V = \sum_{i=1}^{12}(X_i) - 6$$

$$EV = 0$$

$$EX_i = 1/2$$

$$EV^2 = E\sum_{i,j=1}^{12} X_i X_j - 12 E\sum_{i=1}^{12} X_i + 36$$

$$EV^2 = E\sum_{j=1}^{12}\sum_{i=1}^{12}(X_i - 1/2)(X_j - 1/2)$$

$$= E\sum_{i=1}^{12}(X_i - 1/2)^2 = 1$$

where the central limit theorem is used to justify the approximate Gaussianness of V. Note that the tails of the V distribution differ significantly from tails of a true Gaussian distribution. An alternate way of generating Gaussian random variables from uniform random variables that is exact is as follows. Let x_1 and x_2 be independent Gaussian random variables that have a joint distribution function $F(k_1, k_2)$. Then

$$F(k_1, k_2) = \int_{-\infty}^{k_1}\int_{-\infty}^{k_2} \frac{1}{2\pi} e^{-(g_1^2 + g_2^2)/2} dg_1 dg_2$$

$$r = \sqrt{(x_1^2 + x_2^2)}$$

$$\theta = \tan^{-1}\frac{x_1}{x_2}.$$

Changing variables to r and θ, the following expression results:

$$F(R) = \frac{1}{2\pi}\int_0^R e^{-\frac{r^2}{2}} r\,dr\,d\theta.$$

1. Random Noise Generation

$$F(R) = \int_0^R e^{-\frac{r^2}{2}} r \, dr = 1 - e^{-\frac{R^2}{2}}$$

since

$$P(F^{-1}(u) \le a) = P(u \le F(a)) = F(a).$$

When u is uniform, it follows that

$$r = \sqrt{-2\log(1-u_1)}$$

$$\theta = 2\pi u_2$$

are Rayleigh and uniform, respectively, when u_1 and u_2 are uniform. Now it follows that g_1 and g_2 are Gaussian, where

$$g_1 = r\cos(\theta)$$

$$g_2 = r\sin(\theta).$$

Consider again the pseudo random number generator used to produce the uniform random variables, $l_0 =$ seed. Periodicity depends on the seed, and short periods produce poor approximate uniform random variables. For the theoretical underpinnings of random number generators, see [5]. The following **machine-independent** random number generator was introduced by Ken Senne [6]. Let q be the number of pieces into which the random number is broken down for representation on a given machine, then

$$l_{n+1} = a.l_n \pmod{2^{36}}$$

$$q = 3$$

$$l_n = l_n^1 2^{24} + l_n^2 2^{12} + l_n^3 2^0$$

$$a = a^1 2^{24} + a^2 2^{12} + a^3 2^0$$

$$a.l_n = a^1 l_n^1 2^{48} + (l_n^1 a^2 + a^1 l_n^2) 2^{36}$$

$$+ (a^1 l_n^3 + a^3 l_n^1 + a^2 l_n^2) 2^{24} + (l_n^2 a^3 + l_n^3 a^2) 2^{12} + a^3 l_3^n 2^0$$

$$u = \sum_{i=1}^{q-1} l_n^i 2^{-m/q(q-i)}.$$

2 Cholesky Decomposition

Given a symmetric and positive definite matrix A, (a_{nm}), i.e., $A' = A$, and that $x'Ax > 0$ for any nonzero $x \in R^n$. The Cholesky algorithm finds the following factorizations:

$A = U'DU$, where U is an upper triangular matrix whose elements are given by the following relations:

$$a_{i,j}^0 = a_{i,j}$$

$$a_{jk}^i = \left(a_{jk}^{i-1} - \frac{a_{ik}^{i-1} a_{ij}^{i-1}}{a_{ii}^{i-1}} \right), \quad j = i+1, \ldots, n$$

$$l_{ii} = 1$$

$$l_{ij} = \left(\frac{a_{ij}^{i-1}}{a_{ii}^{i-1}} \right), \quad j = i+1, \ldots, n$$

$$d_{ij} = \delta_{i,j} a_{ii}^{i-1}.$$

Since A is positive definite, then

$$\det(A) = \det(D) = \prod_{i=1}^n d_{ii} \leq \prod_{i=1}^n a_{ii}.$$

This inequality, known as Hadamard's theorem, holds with equality iff A is a diagonal matrix; see [24] for a proof. The Cholesky result can be proved by considering the quadratic form induced by A and choosing new variables that depend on successively one less x_i, as

$$\sum_{i=1}^n a_{ij}^0 x_i x_j = a_{11}^0 x_1^2 + 2x_1 \sum_{j=1}^n a_{ij}^0 x_j + \sum_{i=2}^n \sum_{j=2}^n a_{ij}^0 x_i x_j$$

$$= a_{11}^0 (y_1^2) + \sum_{i=2}^n \sum_{j=2}^n \left(a_{ij}^0 - \frac{a_{1i}^0 a_{j1}^0}{a_{11}^0} \right) x_i x_j.$$

Now defining $y_1 = x_1 + \sum_{k=2}^n l_{1j} x_j$ we find that

$$x'Ax = a_{11}^0 y_1^2 + \sum_{i=2}^n \sum_{j=2}^n a_{ij}^1 x_i x_j.$$

Hence our problem has been reduced to diagonalizing a form in $n-1$ variables corresponding to a_{ij}^1 to a weighted sum of squares. Now by iteration we obtain

$$x'Ax = y'Dy$$

$$A = T'DT.$$

This equation indicates how the Cholesky decomposition can be used to generate an arbitrary correlated random vector x from a vector of independent $N(0,1)$ random variables u. Let Σ denote the symmetric positive semidefinite covariance of x. Then by the Cholesky theorem $\Sigma = T'DT$ with T upper triangular and D diagonal with non-negative entries, so that $D^{1/2}TU + m$ is identical in law to x, where m is the mean of x. From the numerical point of view this approach is fast and accurate.

3 Uses of the Pseudo Inverse

For a random vector y with a distribution of the form $N(0, \Xi)$ with $(\Xi) \geq 0$, positive semidefinite, and marginal distributions
$N(0, \Sigma_{xx})$ and $N(0, \Sigma_{yy})$, then the following lemma results.

Lemma 2.1

$$\mathrm{E}x \mid y = \Sigma_{xy}(\Sigma_{yy}^{\#})y. \tag{2.1}$$

Problem 2.1 *Prove Lemma 2.1 is valid when Σ_{yy} is singular.*

The conditional expectation in Lemma 2.1 is expressed via the Moore-Penrose pseudo inverse, which is uniquely defined by the following four axioms:

$$AA^{\#}A = A$$

$$A^{\#}AA^{\#} = A^{\#}$$

$$(AA^{\#})' = AA^{\#}$$

$$(A^{\#}A)' = A^{\#}A.$$

Considering the Moore-Penrose pseudo inverse, we will demonstrate that when the equation $Ax = b$ has a solution it is given by

$$x^0 = A^{\#}b$$

and in general it is an x that minimizes

$$\|Ax - b\|^2.$$

This norm squared is

$$\|A(x - x^0) + Ax^0 - b\|^2$$
$$= \|A(x - x^0)\|^2 + 2\Big(A(x - x^0), Ax^0 - b\Big) + \|Ax^0 - b\|^2.$$

In order to show norm minimization it is sufficient to show that

$$2(x - x^0, A'AA^{\#}b - A'b) = 0.$$

Now, this cross term can be rewritten as

$$(x - x^0, A'(AA^\#)b - A'b) = (x - x^0, A' A^{\#'} A'b - A'b)$$

but this last expression is 0 by the pseudo inverse relation $BB^\# B = B$ for $B = A'$, and hence x^0 is the least-squares solution.

4 Signal Models

Definition 2.1 $\mathbf{x}_t$ $t \in I$ *(the integers), a random process, is Markov iff*

$$P(\mathbf{x}_t \mid \mathbf{x}_s, \; s \leq r) = P(\mathbf{x}_t \mid \mathbf{x}_r) \quad \forall \; r \leq t,$$

where $P(\mathbf{x}_t \mid \mathbf{x}_r)$ *is a transition probability of the Markov process.*

A particular model that generates a Markov process is

$$\mathbf{x}_{n+1} = \phi_{n+1} \mathbf{x}_n + G_{n+1} u_n$$

$$\mathbf{x} \in R^d \quad u \in R^r$$

$$\mathbf{x}_{n_0} = c$$

$$c \in N(0, \Gamma),$$

where $u_n \in N(0, I)$ and $E u_n u_s' = I \delta_{ns}$ and c is independent of the u_n white noise sequence.

Remarks:

1. Gaussian random vectors are closed under linear combinations. This follows from the form of the characteristic function, the Fourier Stieltjes transform of the distribution function. ϕ_{n+1} is the transition matrix for the state from $t = n$ to $t = n + 1$ and is not necessarily nonsingular.

2. By construction, the $\mathbf{x}_t$ sequence is both Gaussian and Markov.

3. $W_{n,m}$, the state covariance satisfies

$$\mathrm{E} x_{n+1} x'_{m+1} = \phi_{n+1}\phi_n \cdots \phi_{m+3}\phi_{m+2} \mathrm{E} x_{m+1} x'_{m+1}$$

for $n \geq m$.

Since the state equation for $\mathbf{x}_{n+1}$ yields

$$x_{n+1} = \Psi_{n+1,m+1} x_{m+1} + \sum_{l=m+1}^{n} K_{n,l} u_l.$$

Let $S_{n+1} = \mathrm{E} x_{n+1} x'_{n+1}$. Then

$$\begin{aligned}
S_{n+1} &= \mathrm{E}(\phi_{n+1} x_n + G_{n+1} u_n)(\phi_{n+1} x_n + G_{n+1} u_n)' \\
&= \Phi_{n+1} S_n \Phi'_{n+1} + G_{n+1} G'_{n+1},
\end{aligned}$$

where the first term on the right-hand side is bilinear and the initial condition is $S_{n_0} = \Gamma$.

Definition 2.2 (ϕ_n, G_n) is (uniformly) completely controllable iff $\forall\, n_0 \ni N(n_0) \geq n_0$ implies $S_N > 0$ for $S_{n_0} = 0$. (There exists an r so that with $N = n_0 + r$ and real numbers α and β, $\beta I \geq S_N \geq \alpha I > 0$ for all n_0.)

The probability distribution for any sequence may be written as

$$P(\mathbf{x}_{t_1}, \ldots, \mathbf{x}_{t_n}) = P(\mathbf{x}_{t_n} \mid \mathbf{x}_{t_1}, \ldots, \mathbf{x}_{t_{n-1}}) P(\mathbf{x}_{t_1}, \ldots, \mathbf{x}_{t_{n-1}}).$$

Use the Markov property, and this becomes

$$= P(\mathbf{x}_{t_n} \mid \mathbf{x}_{t_{n-1}}) P(\mathbf{x}_{t_1}, \ldots, \mathbf{x}_{t_{n-1}}).$$

Hence, the probability distribution for $N(\mu, \Sigma)$ has mean

$$\mu_{n|n-1} = \Psi_{n,n-1} \mathbf{x}_{n-1}$$

$$\sigma_{n|n-1} = G_{n,n-1}.$$

Iteratively apply this to get the joint probability distribution for any pair of sequences. This is the strength of the Markov property.

5 Sensor Model

Now an observation model can be given as

$$z_n \in R^s$$

$$z_n = H_n \mathrm{x}_n + v_n$$

$$E v_n v'_m = \delta_{n,m} R_n, \qquad R_n > 0 \text{ (i.e., positive definite)}.$$

Definition 2.3 (ϕ_n, H_n) *is (uniformly) completely observable if there exists an n such that the map from $z_{n_0}, \ldots, z_n$ to x_{n_0} is invertible, when $u_\bullet = 0$. (When there exists a p so that $n = p + n_0$ and the Jacobian of the map is uniformly bounded above and below for all n_0.)*

Lecture 3

Historical Background

1 Background Material

For statistics references, we recommend [7], [8], and [9]. It is significant that in his book Muirhead [8] is able to find the distribution of the eigenvalues and eigenvectors for real systems of Gaussian random vectors:

$$\mathbf{x}_i \in N(0, \Gamma) \quad ; \quad S = \frac{1}{n} \sum_{i=1}^{n} \mathbf{x}_i \mathbf{x}_i'.$$

For complex Gaussian random variables, see [10]. For hypothesis testing, see [11].

However, the problem with most statistician's work is that "they never think about dynamics (especially nonstationary statistical systems)." The problem of hypothesis testing in the presence of a stochastic process does not appear in any of the above references. This type of hypothesis test tests whether a process, x_s, is present along with the stochastic process, ν_s; more precisely, the testing problem is

$$\begin{cases} H_0 & \nu_s \\ H_1 & x_s + \nu_s \end{cases} \quad s \in (0, t).$$

We will see later that filtering theory resolves this testing problem in the general nonlinear, non-Gaussian case.

2 Historical Developments for Filtering

In 1941, two important ideas for filtering theory came from the study of stationary processes being observed over an infinite interval. These ideas were due to:

- Kolmogorov — his idea was the concept of "innovations".

- Wiener — his idea was to characterize the weighting function of the optimum filter as a solution of the Wiener–Hopf equation.

2.1 Concept of Innovations

Given a sequence of observations (in time sequence):

$$z_i(\omega), \quad i = 1, \ldots, n,$$

where $z_\bullet(\omega)$ is a sample path. What Wiener wanted was the average over all realizations of the process, an ensemble average

$$\int f(z_\bullet(\omega))\, dP(\omega);$$

however, what he had was the time average of the process:

$$\frac{1}{T} \sum_{s=1}^{T} f(z_s(\omega)).$$

The resolution of this dilemma was the ergodic assumption for the process, i.e., that time averages were ensemble averages. Establishing that a given process is ergodic is never easy and is very often simply an ad hoc assumption.

The innovations process is the new information in z_{i+1} that is not in z_i. In other words, it is the part of z_{i+1} that is orthogonal to z_i. First describe the joint information, $I(x, y)$, between two random variables as follows:

$$I(x, y) = \mathrm{E}\left\{ \ln\left(\frac{p(x, y)}{p(x)p(y)} \right) \right\}.$$

2. Historical Developments for Filtering

When x, y are independent, then $I(x,y) = 0$. Creation of the innovations process is similar to the Gram-Schmidt orthogonalization for vectors, as in the Gaussian case independence is equivalent to uncorrelated or orthogonal in the inner product space with inner product $(x,y) = Exy$. The orthogonalization steps are as follows:

$$I_1 = z_1$$

$$I_2 = z_2 - \mathrm{E}\left\{z_2 z_1'\right\} \left(\mathrm{E}\left\{z_1 z_1'\right\}\right)^{-1} z_1$$

or by Lemma 2.1,

$$I_2 = z_2 - \mathrm{E}\{z_2 | z_1\}.$$

Note that I_2 is independent of I_1, since they are uncorrelated by construction.

<u>Motivation:</u> Kolmogorov knew that if he had two independent random variables, say x_1 and x_2 with variance σ_1 and σ_2, respectively, and equal mean values, $\mathrm{E}(x_1) = \mu$ and $\mathrm{E}(x_2) = \mu$, then one can combine the two random variables so as to minimize

$$\mathrm{E}\left\{(\mu - \alpha x_1 - \beta x_2)^2\right\} = \tau = \mathrm{E}\left\{\alpha^2(\mu - x_1)^2\right\} + \mathrm{E}\left\{\beta^2(\mu - x_2)^2\right\}.$$

In this equation, to have an unbiased estimate requires $\alpha + \beta = 1$ from the relation $\mathrm{E}(\alpha x_1 + \beta x_2) = 2\mu$. Hence,

$$\tau = (1-\beta)^2 \sigma_1^2 + \beta^2 \sigma_2^2 + 2(1-\beta)\beta\sigma_{12}^2,$$

where $\sigma_{12}^2 = 0$ for this case since x_1 and x_2 are assumed to be independent. Now, differentiate with respect to β and set to zero to find the minimum:

$$-2(1-\beta)\sigma_1^2 + 2\beta\sigma_2^2 = 0 \quad \Rightarrow$$

$$\beta = \frac{\sigma_1^2}{\sigma_1^2 + \sigma_2^2}, \quad \alpha = \frac{\sigma_2^2}{\sigma_1^2 + \sigma_2^2} \tag{3.1}$$

so that

$$\hat{\mu} = \frac{\sigma_2^2 x_1}{\sigma_1^2 + \sigma_2^2} + \frac{\sigma_1^2 x_2}{\sigma_1^2 + \sigma_2^2}.$$

This can be extended to n measurements by finding the common denominator and rewriting as

$$\hat{\mu} = \frac{\frac{x_1}{\sigma_1^2} + \frac{x_2}{\sigma_2^2}}{\frac{1}{\sigma_1^2} + \frac{1}{\sigma_2^2}}.$$

However, if x_1 and x_2 are not independent, then instead of Equation (3.1), the equation to find the minimum β becomes

$$2(1-\beta)\sigma_1^2 = 2\beta\sigma_2^2 + (1-2\beta)\sigma_{12}^2 \quad \Rightarrow \quad \beta = \frac{\sigma_1^2 + \sigma_{12}^2}{\sigma_1^2 + \sigma_2^2 + 2\sigma_{12}^2};$$

hence, in this case the simple extension to the case of n random variables is not possible as in the previous case.

Problem 3.1 *Do the above calculation for three measurements. These equations should be familiar as the equivalent parallel resistance from circuit theory. This illustrates the ease with which independent variables can be combined for estimates of their (common) mean.*

As an aside, note that Duffin generalized resistance. The preceding for matrices is

$$A(A+B)^{\#} B.$$

2.2 Wiener-Hopf Equation

Lemma 3.1 *Let x, y, and z be zero mean Gaussian random variables. Then if y and z are independent,*

$$E\{x|y,z\} = E\{x|y\} + E\{x|z\}. \tag{3.2}$$

2. Historical Developments for Filtering

Proof:

$$LHS \sim E\left\{ x \begin{pmatrix} y \\ z \end{pmatrix}' \left[E \begin{pmatrix} y \\ z \end{pmatrix} \begin{pmatrix} y \\ z \end{pmatrix}' \right]^{-1} \begin{pmatrix} y \\ z \end{pmatrix} \right\}$$

$$= \begin{pmatrix} E\{xy'\} \\ E\{xz'\} \end{pmatrix}' \begin{bmatrix} \Sigma_y & 0 \\ 0 & \Sigma_z \end{bmatrix}^{-1} \begin{pmatrix} y \\ z \end{pmatrix}$$

$$= E\{xy'\} \Sigma_{yy}^{-1} + E\{xz'\} \Sigma_{zz}^{-1},$$

which is the RHS. QED

We are interested in

$$E\{x_t | z_1, \ldots, z_n\}$$

because we want to estimate x_t (the signal) from z_i (observations). The terminology for the possibilities for this type of estimation problem will be defined as follows:

- if $t > n$ extrapolation

- if $t < n$ interpolation

- if $t = n$ filtering or smoothing.

This whole problem was motivated in the 1940s by the need for real-time fire-control problems in World War II. Wiener posed the problem as follows: $x_t \in R^d$ is the signal and z_i, $i = 1, \ldots, n$ are observations. Construct the estimate, $\hat{x}_t$, which

1. is unbiased such that $E\{\hat{x}_t\} = E\{x_t\} = 0$

2. has minimum error variance: $E[\lambda'(x_t - \hat{x}_t)]^2 = \min \ \forall \ \lambda \in R^d$.

We will show that $\hat{x}_t = E(x_t|z_1,\ldots,z_t)$.

To show that the estimate $\hat{x}_t$ has minimum variance, say there is some other unbiased estimator, $x_t^* = F(z_1,\ldots,z_t)$, so that $E(\lambda x^*)^2 < \infty$ for all λ.

$$E\{\lambda'(x_t - x_t^*)\}^2 = E\{\lambda'[(x_t - \hat{x}_t) + (\hat{x}_t - x_t^*)]\}^2 =$$

$$E\{\lambda'(x_t - \hat{x}_t)\}^2 + 2E\left\{\underbrace{\lambda'(x_t - \hat{x}_t)}_{G_1}\overbrace{\lambda'(\hat{x}_t - x_t^*)}^{G_2}\right\} + E\{\lambda'(\hat{x}_t - x_t^*)\}^2$$

so that now we have only to show $E\{G_1 G_2\} = 0$ for $\hat{x}_t$ to have minimum variance, i.e., choose $\hat{x}_t = x_t^*$. Hence,

$$E\{G_1 G_2\} = E\{G_1 G_2 | z_1,\ldots,z_n\}$$

$$= EG_2\{EG_1 | z_1,\ldots,z_n\},$$

where we have used the theorem that says that if G_2 is measurable with respect to the conditioning field, then we can pull it out of the expectation as shown earlier. But from the definition of $\hat{x}$ we have that

$$E\{G_1 | z_1,\ldots,z_n\} = 0. \qquad \text{QED}$$

Furthermore, if $x^* = 0$ then the cross term becomes

$$E\{\lambda'(x_t - \hat{x}_t)\lambda'(\hat{x}_t)\} = 0.$$

We could now show (as was done earlier to show that $\hat{x}_t$ was the minimum variance estimate) that the Wiener-Hopf equation can be written as

$$E\{x_t F'(z_1,\ldots,z_n)\} = E\{\hat{x}_t F'(z_1,\ldots,z_n)\}$$

2. Historical Developments for Filtering

by a property of the conditional expectation, namely,

$$Efg|\mathcal{O} = fEg|\mathcal{O}$$

for $f\mathcal{O}$ measurable so that by the smoothing property

$$Efg = Ef(Eg|\mathcal{O}).$$

Now letting $g = x_t$, $f = F'(z_1, \ldots, z_n)$ with $\mathcal{O}$ the minimal σ-field over the z_i, $i = 1, \ldots, n$, the Wiener-Hopf equation follows.

Wiener assumed the following:

$$\hat{x}_t = \int_{-\infty}^{t} w(t-s) z(s) \, ds,$$

so that taking $F = z'(\tau)$ we find that

$$E\{x(t) z'(\tau)\} = \int_{-\infty}^{t} w(t-s) E\{z(s) z'(\tau)\} \, ds \quad \forall \, \tau \leq t.$$

This is the Wiener-Hopf equation for a continuous time system.

Wiener's scalar stationary assumption, $z = x + \nu$, where ν is white noise with spectral density N_0, results in a Fredholm integral equation of the second kind:

$$\Phi(t-\tau) = \int_{-\infty}^{t} w(t-s) \Phi(s-\tau) \, ds + N_0 w(t-\tau); \quad \forall \, \tau \leq t, \quad (3.3)$$

where

$$\Phi(t) = E\{x(t)x(0)\}.$$

<u>Result No. 1</u>: $\hat{x}_t = E\{x_t | z_1, \ldots, z_n\}$

<u>Result No. 2</u>: $E\{x_t z_s'\} = E\{\hat{x}_t z_s'\}, \quad \forall \, 1 \leq s \leq n.$

For the linear Gaussian problem this is all that is needed since Result

No. 1 implies that $\hat{x}$ is a linear combination of $z_i's$ and Result No. 2 gives the coefficients of this linear combination. If $n = t$, the conditional distribution of x_t is as follows:

$$N\left(\hat{x}_t; \Sigma_{x_t x_t} - \Sigma_{x_t z^t} \Sigma_{z^t z^t}^{-1} \Sigma_{z^t x_t}\right),$$

where the second term in parentheses is the variance of $\hat{x}_t$ and $z^t = (z_1, \ldots, z_t)'$. However, this is not a practically effective way to calculate the variance for large numbers of observations. For this, we return to consideration of the innovations.

3 Development of Innovations

Define the sequence of innovations as follows:

$$I_1 = z_1$$

$$I_2 = z_2 - C_2{}^1 I_1$$

$$I_3 = z_3 - \sum_{j=1}^{2} C_3{}^j I_j$$

$$\vdots$$

$$I_n = z_n - \sum_{j=1}^{n-1} C_n{}^j I_j.$$

Now these equations imply that, for all measurements, the relation between innovations and measurements takes the form of a lower block triangular matrix:

$$z^n = \begin{bmatrix} I & 0 & 0 & \cdots & & 0 \\ C_2^1 & I & 0 & \cdots & & 0 \\ C_3^1 & C_3^2 & I & 0 & \cdots & 0 \\ \vdots & \vdots & & \ddots & & \\ C_n^1 & C_n^2 & \cdots & & & I \end{bmatrix} I^n. \tag{3.4}$$

4. Sequential Filter Development

Theorem 3.1 $f : z^n \to I^n$ *is bijective.*

<u>Proof:</u> Recall that v^n is the column vector $(v_1, \ldots, v_n)'$. Since the matrix relating z^n to I^n is lower block triangular it is invertible.

Remark 3.1 *In continuous time with Gaussian white noise this theorem has evaded efforts at proof in the nonlinear case.*

Corollary 3.1

$$\mathcal{F}(z_1, \ldots, z_n) = \mathcal{F}(I_1, \ldots, I_n) \quad (\mathcal{F} \text{ denotes } \sigma \text{ field}).$$

In particular, conditional expectations are the same with respect to either $z_\bullet$ or $I_\bullet$.

Corollary 3.2

$$\begin{aligned} \mathrm{E}\{x_t | z_1, \ldots, z_n\} &= \mathrm{E}\{x_t | I_1, \ldots, I_n\} \\ &= \sum_{j=1}^{n} \mathrm{E}\{x_t I_j'\} \mathrm{E}\{I_j I_j'\}^{-1} I_j. \end{aligned}$$

<u>Proof:</u> See Lemma 3.1.

Note that the relation is valid for both $t \geq n$ and $t < n$, so that it provides a solution to the smoothing, extrapolation, and interpolation problems. The Is are not normalized and have different variances. Note that this is not a sequential algorithm since all the Is must be known at each step.

4 Sequential Filter Development

The steps that follow will outline the derivation of a sequential filter. Historically, Kalman considered the prediction problem without observation noise in his 1960 ASME paper. Here we will develop the full Kalman-Bucy filter including plant noise and observation noise. The original Kalman derivation can be obtained by assuming the observation noise is zero in what follows. The system equation and observation equation are represented as

$$x_{t+1} = \Phi_{t+1} x_t + G_{t+1} u_t, \tag{3.5}$$

$$z_t = H_t x_t + \nu_t, \tag{3.6}$$

where the first vector equation represents the state, x_{t+1}, at time $t+1$ in terms of the state at time t and the "plant noise", u_t, which will be assumed to have known variance $\mathrm{E}\{u_t u_s'\} = Q_s \delta_{t,s}$. Further assume that the plant noise has zero mean, $\mathrm{E}\{u_t\} = 0$. The second vector equation represents the observation process with "observation noise", ν_t, which will be assumed to have known variance $\mathrm{E}\{\nu_t \nu_s'\} = R_s \delta_{t,s}$. Here we define the following cases:

- the prediction problem — implies there is no observation noise.
- the parameter estimation problem — (also called least squares) implies there is no plant noise.

Now define the one-step predictor and the filter as follows:

$$\hat{x}_{t+1|t} \triangleq \mathrm{E}\{x_{t+1} \mid z_1, \ldots, z_t\} \quad (1-\text{step predictor})$$

$$\hat{x}_{t|t} \triangleq \mathrm{E}\{x_t \mid z_1, \ldots, z_t\} \quad (\text{filter}).$$

To show that these relations are in fact correct, first prove the linear state estimate mapping (there will be a nonlinear analog later in the book).

Theorem 3.2 *The filter state estimate is related to the one-step predictor as follows:*

$$\hat{x}_{t+1|t} = \Phi_{t+1} \hat{x}_{t|t}. \tag{3.7}$$

Proof: It suffices to show that

$$\mathrm{E}\{u_t \mid z_1, \ldots, z_t\} = 0,$$

4. Sequential Filter Development

but, this is implied by

$$E\{u_t\} = 0$$

since u_t depends only on $z_{t-1}, \ldots, z_1$ and by the assumption of zero mean plant noise. This is the case in view of the recurrence relation defining x_{t+1} in terms of x_t and u_t; see Equation (3.5).

Furthermore, the covariance of the filter error is given by

$$P_t = E\left\{(x_t - \hat{x}_{t|t-1})(x_t - \hat{x}_{t|t-1})'\right\}. \tag{3.8}$$

Theorem 3.3 *The filter estimate is related to the innovations as follows:*

$$\hat{x}_{t|t} = \hat{x}_{t|t-1} + K_t(I_t). \tag{3.9}$$

Proof: Since I_t and $z_1, \ldots, z_{t-1}$ are independent, one can write

$$E\{x_t|z_1, \ldots, z_t\} = E\{x_t|z_1, \ldots, z_{t-1}, I_t\}$$

$$= E\{x_t|z_1, \ldots, z_{t-1}\} + E\{x_t|I_t\}$$

$$= \hat{x}_{t|t-1} + K_t(I_t).$$

Corollary 3.3

$$K_t = E\{x_t I_t'\}\left(E\{I_t I_t'\}\right)^{-1}. \tag{3.10}$$

Lemma 3.2

$$I_t = z_t - E\{x_t|z_1, \ldots, z_{t-1}\}.$$

Proof: $I_t = H\hat{x}_{t|t-1} + \nu_t$ and $EI_t z_s' = HE\hat{x}_{t|t-1}z_s' = 0$ by the Wiener-Hopf equation.

Theorem 3.4 *The sequential filter is given by*

$$\hat{x}_{t+1/t} = \Phi_{t+1}\hat{x}_{t/t-1} + \Phi_{t+1}K_t\left(z_t - H\hat{x}_{t/t-1}\right), \quad (3.11)$$

$$\hat{x}_{1/0} = 0, \quad (3.12)$$

$$K_t = P_t H_t'(H_t P_t H_t' + R_t)^{-1}. \quad (3.13)$$

Proof: This is a consequence of Theorems 3.2 and 3.3 and the following error variance relations:

$$\tilde{x}_{t/t-1} = x_t - \hat{x}_{t/t-1}$$

$$E\{\tilde{x}_{t/t-1} z_s'\} = 0 \Rightarrow$$

$$E\{\tilde{x}_{t/t-1} \hat{x}_{t/t-1}'\} = 0$$

$$\Rightarrow P_t = E\{\tilde{x}_{t/t-1} x_t'\},$$

where the matrix inverse in Equation (3.13) was originally the Moore-Penrose pseudo inverse in Kalman's 1960 paper, since he assumed $R_t = 0$.

In view of the last theorem and the dynamical equation for the state,

$$\tilde{x}_{t+1/t} = \Phi_{t+1}\left(I - K_t H_t\right)\tilde{x}_{t/t-1} + G_{t+1}Q_t + \Phi_{t+1}K_t\nu_t.$$

If this equation is cross correlated with the state equation, we obtain the covariance

$$P_{t+1} = G_{t+1}Q_t G_{t+1}' + \Phi_{t+1}\left(P_t - K_t H_t P_t\right)\Phi_{t+1}' \quad (3.14)$$

or

$$P_{t+1} = \Phi_{t+1}\left(P_t - P_t H_t'(H_t P_t H_t' + R_t)^{-1} H_t P_t\right)\Phi_{t+1}' + G_{t+1}Q_t G_{t+1}',$$

4. Sequential Filter Development

which is the matrix Riccati equation for the linear, nonstationary, multi-input and multioutput filter. Now, to get the historic Kalman result, let $R_t \to 0$ and make all the coefficient matrices constant. This gives

$$P_{t+1} = \Phi \left[P_t - P_t H'(HP_t H')^{\#} HP_t \right] \Phi' + C,$$

where the term in brackets is

$$\Sigma_t = P_t - P_t H'(HP_t H')^{\#} HP_t .$$

Note that by the lemma of the pseudo inverse operation

$$\Sigma_t H' = 0.$$

This is an indication of the "stiffness" in the predictor; we therefore cannot guarantee stability. This is the ill-conditionedness of the least-squares solution.

Lecture 4

Sequential Filtering Theory

1 Summary of the Sequential Filter

Given the following model for the d-dimensional system and observations:

$$x_{n+1} = \Phi_{n+1} x_n + G_{n+1} u_n$$

$$z_n = H_n x_n + \nu_n,$$

where $x_1 \in N(0, \Gamma)$ and

$$E u_n u'_m = \delta_{n,m} Q_n$$

$$E \nu_n \nu'_m = \delta_{n,m} R_n$$

and $Q_n \geq 0$. We discovered last time that the sequential filter is described by the equations

$$\hat{x}_{n+1|n} = \Phi_{n+1} \hat{x}_{n|n-1} + K_n(I_n), \tag{4.1}$$

$$I_n = z_n - H_n \hat{x}_{n|n-1}, \tag{4.2}$$

$$K_n = \Phi_{n+1} P_n H'_n (H_n P_n H'_n + R_n)^{-1}, \tag{4.3}$$

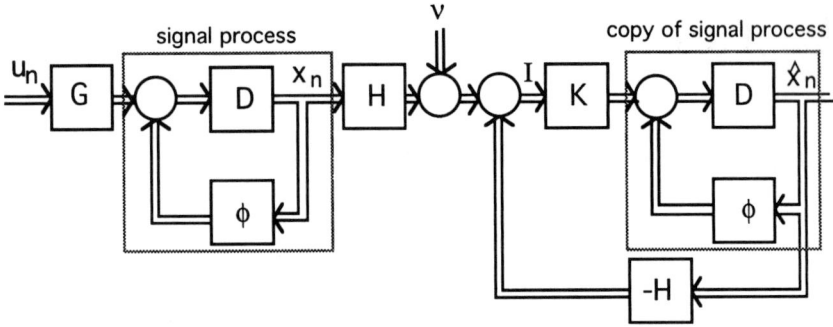

Figure 4.1: System diagram for the linear filter.

$$P_{n+1} = \Phi_{n+1}\left(P_n - P_n H_n'(H_n P_n H_n' + R_n)^{-1} H_n P_n\right)\Phi_{n+1}'$$

$$+ G_{n+1} Q_n G_{n+1}', \qquad (4.4)$$

where $\hat{x}_{1|0} = 0$ and P_n is the error covariance matrix, defined by

$$P_n = E\tilde{x}_{n|n-1}\tilde{x}'_{n|n-1}$$

$$\tilde{x}_{n|n-1} = x_n - \hat{x}_{n|n-1}$$

so that $P_0 = \Gamma$. $\tilde{x}_{n|n-1}$ is called the filter error. Equation (4.4) is the matrix Riccati equation for the filter.

The system, consisting of the model plus the filter, can be represented by the block diagram shown in Figure 4.1. The block denoted D in Figure 4.1 represents a unit delay; the other blocks are related to the matrices given in equations (4.1) and (4.3). Note that the filter contains a copy of the signal process. While the input to the real signal process is u_n, which is a white noise process, the input to the copy within the filter is $K_n I_n$, which is also white. Note that I_n is not white in general: it has covariance $H_n P_n H_n' + R$. The gain K_n is chosen to turn I_n into a white noise process. In an optimal filter, $K_n I_n$ is white.

2. The Scalar Autonomous Riccati Equation

Remark: Dr. James Follin [16] has pointed out that if u_n is moved over to the copy of the signal process within the filter, the output of the filter is $\tilde{x}$ instead of $\hat{x}$.

Remark: Note that the Markov property is essential in the derivation of the filter since the system equation is intrinsically Markov:

$$P(x_{n+1}, x_n, \ldots, x_1) = P(x_{n+1} \mid x_n) P(x_1, \ldots, x_n).$$

Remark: Note that the prediction (i.e., extrapolation) problem has also been solved, since

$$\hat{x}_{n+\tau|n} = \Phi_{n+\tau} \Phi_{n+\tau-1} \cdots \Phi_{n+2} \hat{x}_{n+1|n},$$

where $\tau > 1$.

2 The Scalar Autonomous Riccati Equation

We want to consider the scalar Riccati equation and prove some asymptotic stability results. Let $\Phi = 1$, $R = r$, $Q = q$, and $G = H = 1$ in Equation (4.4). The Riccati equation then becomes

$$p_{n+1} = p_n - p_n(p_n + r)^{-1} p_n + q$$

$$p_1 = \gamma > 0.$$

Simplifying this, we get

$$p_{n+1} = \frac{p_n r}{p_n + r} + q.$$

At this point, we could let $r = 1$ without loss of generality, in which case p_n would be expressed in the form $r f(q/r)$; it is the ratio q/r that is important. However, we will continue with $r \neq 1$.

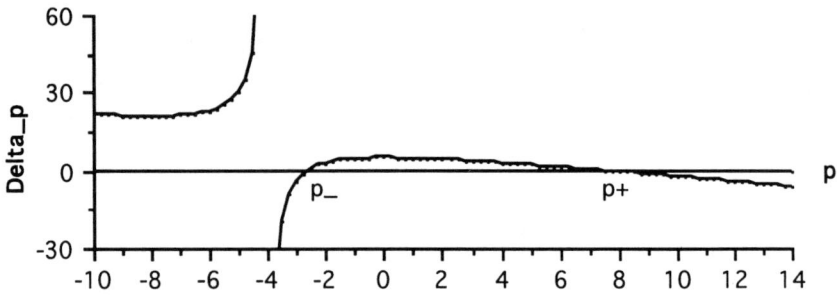

Figure 4.2: Phase diagram for scalar Riccati equation.

Rearranging the preceding equation, we get

$$\Delta p = p_{n+1} - p_n = q - \frac{p_n^2}{p_n + r}.$$

We can now draw a phase diagram for p as shown in Figure 4.2. For drawing the figure, values of $r = 4$, $q = 5$ have been assumed. The equilibrium points for p occur when $\Delta p = 0$, i.e., when

$$p^2 - qp - qr = 0.$$

Thus there are two equilibrium points:

$$p_+ = \frac{q}{2} + \sqrt{\frac{q^2}{4} + qr}$$

$$p_- = \frac{q}{2} - \sqrt{\frac{q^2}{4} + qr}.$$

Provided $q > 0$ and $r > 0$, the sign of Δp is given by

$$\Delta p < 0 \quad \text{if} \quad p > p_+$$

$$\Delta p > 0 \quad \text{if} \quad p_- < p < p_+$$

2. The Scalar Autonomous Riccati Equation

$$\Delta p < 0 \quad \text{if} \quad -r < p < p_-$$

$$\Delta p > 0 \quad \text{if} \quad p < -r.$$

<u>Conclusion:</u> As long as $q, r > 0$, $\Pi(n, \gamma) \longrightarrow p_+$ as $n \to \infty$ $\forall \gamma > p_-$.

Let's investigate how quickly p_n approaches p_+. First, note that

$$p_+ = \frac{p_+ r}{p_+ + r} + q.$$

Now define

$$\begin{aligned}
\delta_{n+1} &= p_{n+1} - p_+ \\
&= \frac{p_n r}{p_n + r} - \frac{p_+ r}{p_+ + r} \\
&= \frac{r^2 \delta_n}{(\delta_n + p_+ + r)(p_+ + r)}.
\end{aligned}$$

Now let

$$k_{n+1} = \frac{1}{\delta_{n+1}} = \frac{p_+ + r}{r^2}\left(1 + (p_+ + r)k_n\right).$$

Then k_{n+1} can be expressed in the form

$$k_{n+1} = \alpha k_n + \beta,$$

where

$$\alpha = \left(1 + \frac{p_+}{r}\right)^2$$

$$\beta = \frac{\left(\frac{p_+}{r} + 1\right)}{r}.$$

The general solution to this difference equation is

$$k_{n+1} = \alpha^n k_1 + \sum_{j=1}^{n} \alpha^{n-j} \beta = \alpha^n k_1 + \beta\left(\frac{\alpha^n - 1}{\alpha - 1}\right).$$

Thus we have the general solution for δ_{n+1}:

$$\delta_{n+1} = \frac{1}{\alpha^n k_1 + \frac{\beta(\alpha^n - 1)}{\alpha - 1}}$$

$$= \frac{\delta_1/\alpha^n}{1 + \frac{\beta \delta_1}{\alpha - 1}\left(1 - \frac{1}{\alpha^n}\right)}.$$

Thus, δ_n decreases as $\left(1/1 + \frac{p_+}{r}\right)^n$, which implies exponential convergence.

<u>Further questions</u>: How long does it take to forget the initial conditions? What happens when α gets very large?

Next, let's consider what happens when $-r < p < p_-$. Since $\Delta p < 0$, the sequence of points p_n will move away from the equilibrium point towards $-r$. If p_n should land exactly on $-r$, the Riccati equation blows up, since p_{n+1} then becomes undefined. In fact, there is a countable number of points $\{\ldots, r_n, \ldots, r_2, r_1\}$ in the interval $-\infty < p < p_-$, all of which eventually reach $-r$ and cause the Riccati equation to blow up. These are called *conjugate points*. Let's find the first of these, r_1, which is defined so that $p_{n+1} = -r$ when $p_n = r_1$. We first rearrange the Riccati equation to express p_n in terms of p_{n+1}:

$$p_n = \frac{r(q - p_{n+1})}{p_{n+1} - r - q}.$$

Substituting $p_{n+1} = -r$, we get

$$r_1 = \frac{-r(q + r)}{2r + q}.$$

Repeated substitutions give us the whole sequence of conjugate points.

3. Linearizing the Riccati Equation

What happens to points between the conjugate points? Those points in the interval (r_k, r_{k+1}) map to the next interval to the left, (r_{k-1}, r_k), and points in the last interval $(-r, r_1)$ jump to the left of $-r$. Now consider a point to the left of $-r$, at $-r - \zeta$, where $\zeta > 0$. It will map to

$$\frac{r(r+\zeta)}{\zeta} + q > 0.$$

Thus, points between the conjugate points eventually map into the stable region (p_-, p_+).

An understanding of the stability of the discrete Riccati equation is invaluable in simulating continuous systems properly. Our analysis has shown that if you start to the left of p_-, the sequence can explode to $-\infty$ in finite time, although you may not see this because the sequence can end up back in the stable region if points between conjugate points are included. Thus, numerical simulations can easily mask the presence of the conjugate points. This is closely related to the Jacobi accessory problem; see [17].

3 Linearizing the Riccati Equation

Define x_n and y_n such that

$$p_n = \frac{y_n}{x_n}.$$

Then the Riccati equation becomes

$$\frac{y_{n+1}}{x_{n+1}} = \frac{\frac{y_n}{x_n} r}{\frac{y_n}{x_n} + r} + q$$

$$= \frac{y_n r}{y_n + r x_n} + q$$

$$= \frac{y_n + q\left(\frac{1}{r} y_n + x_n\right)}{\frac{1}{r} y_n + x_n}.$$

This is associated with the following linear system:

$$\begin{pmatrix} x_{n+1} \\ y_{n+1} \end{pmatrix} = \begin{pmatrix} 1 & \frac{1}{r} \\ q & 1+\frac{q}{r} \end{pmatrix} \begin{pmatrix} x_n \\ y_n \end{pmatrix},$$

and since this system is linear, it has a solution of the form

$$\begin{pmatrix} x_{n+1} \\ y_{n+1} \end{pmatrix} = S^n \begin{pmatrix} x_1 \\ y_1 \end{pmatrix},$$

where S is *symplectic* (see below), and its determinant is 1. This technique is due to Count Riccati for the continuous-time Riccati equation. The method may also be applied to the multidimensional problem and the associated matrix Riccati equation. Thus we can in general linearize the Riccati equation by looking at a linear equation in a higher-dimensional space.

3.1 Symplectic Matrices

Definition 4.1 *A square matrix A is symplectic iff $A'JA = J$, where*

$$J = \begin{pmatrix} 0 & I \\ -I & 0 \end{pmatrix}.$$

Note that an order 2 matrix with determinant equal to 1 is always symplectic. Symplectic matrices describe the behavior of Hamiltonian systems (see [18]).

Theorem 4.1 *Symplectic matrices form a group.*

Proof: Let $\mathcal{S}$ be the set of symplectic matrices.
Identity: By inspection, $I \in \mathcal{S}$.
Closure: If $A \in \mathcal{S}$ and $B \in \mathcal{S}$, then

$$(AB)' J AB = B'A'JAB = B'JB = J,$$

so $(AB) \in \mathcal{S}$.

3. Linearizing the Riccati Equation

Inverse: If $A \in \mathcal{S}$, then

$$A'JA = J$$

$$(A')^{-1}A'JAA^{-1} = (A^{-1})'JA^{-1}$$

and therefore $A^{-1} \in \mathcal{S}$. We can also show that $A' \in \mathcal{S}$, as follows:

$$JA'JA = JJ = -I$$

$$AJA'JA = -A$$

$$AJA'J = -I$$

$$AJA'JJ' = -J' = J$$

$$AJA' = J.$$

3.2 Stability of the Filter

Consider the estimate update equation for the scalar filter:

$$\hat{x}_{n+1} = \hat{x}_n + k_n(z_n - \hat{x}_n),$$

where

$$k_n = \frac{p_n}{p_n + r}.$$

Let's look at the error filter:

$$\tilde{x}_{n+1} = x_{n+1} - \hat{x}_{n+1}$$

$$= x_n + u_n - \hat{x}_{n+1}$$

$$= (1 - k_n)\tilde{x}_n + \text{random terms}.$$

We can show that the undriven error equation is asymptotically stable. After N steps, the error is given by

$$\tilde{x}_N = \prod_{n=0}^{N-1} \frac{1}{1+p_n/r} \tilde{x}_0 + \text{random terms}.$$

Provided $q > 0$ and $r > 0$, the product will tend to zero after a while, since $p_n \to p_+$, and

$$\frac{1}{1+p_+/r} < 1.$$

This proves the uniform asymptotic stability of the filter.

Remark: There is a relationship between the Riccati equation and continued fractions; see [12], [13], and [14].

Remark: The Riccati equation can be applied in solving the Weiner-Hopf equation; see [15].

Lecture 5

Burg Technique

1 Background Material

The original linear prediction problem with *no* observation noise solved by Norbert Wiener [19] was given a sequential solution for the discrete time for the scalar stationary ergodic observation case by Levinson; see the appendix to [19]. Kalman solved the no observation noise discrete time problem, dropping the stationarity, ergodicity, and scalar assumptions and estimating the entire state; see [33]. Burg [21], under the Levinson assumptions, introduced a novel method for sequential covariance estimation. Burg's method is widely used in that it produces a *stable* output predictor as well as a spectral estimator using just the observations. A stationary process has $\boldsymbol{\Phi}$, $\mathbf{H}$, $\mathbf{G}$, and $\mathbf{Q}$ independent of n. Consider the complex vector sequence $\mathbf{y}_n$ which are stationary, complex d-vectors. Another way of saying this is that

$$E\mathbf{y}_n\mathbf{y}_m^* = \begin{cases} \mathbf{R}(n-m) & \text{if } n \geq m; \\ \mathbf{R}^*(m-n) & \text{otherwise.} \end{cases}$$

The $*$ symbol indicates the hermitian conjugate, and $\mathbf{R}$ is a $d \times d$ matrix.

The prediction problem is to find the conditional expectation

$$E\mathbf{y}_{n+1}|\mathbf{y}_0\ldots\mathbf{y}_n = \hat{\mathbf{y}}_{n+1}$$

($\mathbf{y} = \mathbf{Hx}$, and $\mathbf{z} = \mathbf{y} + \nu$). The Wiener-Hopf equation (see Lecture 3, page

27) says that the error is perpendicular to the data, i.e.,

$$E\tilde{y}_{n+1}y_j^* = 0, \quad j = 0, 1, \ldots n.$$

Burg looks at the $d = 1$ case for which $Ey_n\bar{y}_m = r(n - m)$. We know that after n observations

$$\tilde{y}_{n+1} = \sum_{i=0}^{n+1} \alpha_i^n y_i,$$

where α^n is the set of scalar coefficients determining the predictor. This structure is valid because of the Gaussian distribution of the y_i. In view of the Wiener-Hopf equation it follows that

$$\sum_{i=0}^{n+1} \alpha_i^n Ey_i\bar{y}_j = 0, \quad j = 0, \ldots n.$$

Notice that

$$E\tilde{y}_{n+1}\bar{y}_{n+1} = E|\tilde{y}_{n+1}|^2$$

(see Lecture 3) so that

$$\Rightarrow \sum_{i=0}^{n+1} \alpha_i^n Ey_i\bar{y}_{n+1} = P_{n+1}.$$

This can be expressed as

$$\mathbf{R}_{n+1}\alpha^n = \begin{pmatrix} r_0 & r_{-1} & \cdots & \cdots & r_{-(n+1)} \\ r_1 & r_0 & \cdots & \cdots & r_{-n} \\ \vdots & \vdots & \ddots & & \vdots \\ \vdots & \vdots & & \ddots & \vdots \\ r_{n+1} & r_n & \cdots & \cdots & r_0 \end{pmatrix} \begin{pmatrix} 1 \\ \alpha_n^n \\ \alpha_{n-1}^n \\ \vdots \\ \alpha_0^n \end{pmatrix} = \begin{pmatrix} P_{n+1} \\ 0 \\ 0 \\ \vdots \\ 0 \end{pmatrix}.$$

Remark: $r_{-k} = \bar{r}_k$.

Definition: $\mathbf{A}$ is *Toeplitz* iff $\mathbf{A} = \{a_{ij}\}$, where $a_{ij} = f(i - j)$.

1. Background Material

(N.B. A stationary process has a Toeplitz covariance.) Let

$$\mathbf{J} = \begin{pmatrix} 0 & & & 1 \\ & & 1 & \\ & 1 & & \\ & 1 & & \\ 1 & & & 0 \end{pmatrix}$$

and be of the same dimension as $\mathbf{R}_{n+1}$. This is the reversal matrix: $\mathbf{J}^2 = \mathbf{I}$. Premultiplying by $\mathbf{J}$ reverses rows, and post-multiplying reverses columns.

$$\mathbf{J}\mathbf{R}_{n+1}\mathbf{J} = \bar{\mathbf{R}}_{n+1}$$

$$\mathbf{J}\mathbf{R}_{n+1}\mathbf{J}\mathbf{J}\alpha^n = \mathbf{J}\begin{pmatrix} P_{n+1} \\ 0 \\ \vdots \\ 0 \end{pmatrix}$$

$$\bar{\mathbf{R}}_{n+1}\mathbf{J}\alpha^n = \begin{pmatrix} 0 \\ 0 \\ \vdots \\ P_{n+1} \end{pmatrix}$$

$$\mathbf{R}_{n+1}\mathbf{J}\bar{\alpha}^n = \begin{pmatrix} 0 \\ \vdots \\ 0 \\ P_{n+1} \end{pmatrix}.$$

Suppose on the previous step

$$\mathbf{R}_n \beta^{n-1} = \begin{pmatrix} P_n \\ 0 \\ \vdots \\ 0 \end{pmatrix}.$$

Then

$$\mathbf{R}_n \mathbf{J}\bar{\beta}^{n-1} = \begin{pmatrix} 0 \\ \vdots \\ 0 \\ P_n \end{pmatrix}.$$

Lecture 5. Burg Technique

Remark:

$$\mathbf{R}_{n+1} = \begin{pmatrix} r_0 & r_{-1} & \cdots & \cdots \\ r_1 & & & \\ \vdots & & \mathbf{R}_n & \\ \vdots & & & \end{pmatrix} \tag{5.1}$$

$$= \begin{pmatrix} & & & r_{-(n+1)} \\ & \mathbf{R}_n & & \vdots \\ & & & \vdots \\ r_{n+1} & \cdots & \cdots & r_0 \end{pmatrix}. \tag{5.2}$$

Assuming $\det(\mathbf{R}_n) \neq 0$, we will construct α^n from β^{n-1}s.

$$\mathbf{R}_{n+1} \begin{pmatrix} \beta^{n-1} \\ 0 \end{pmatrix} = \begin{pmatrix} P_n \\ 0 \\ \vdots \\ 0 \\ \mathbf{z}'\beta^{n-1} \end{pmatrix},$$

where $\mathbf{z}' = (r_{n+1} \ r_n \ \cdots \ r_1)$ using the second representation of $\mathbf{R}_{n+1}$, Equation (5.2). Also look at

$$\mathbf{R}_{n+1} \begin{pmatrix} 0 \\ \mathbf{J}\bar{\beta}^{n-1} \end{pmatrix} = \begin{pmatrix} \bar{\mathbf{z}}'\bar{\beta}^{n-1} \\ 0 \\ \vdots \\ 0 \\ P_n \end{pmatrix}.$$

Then take

$$\mathbf{R}_{n+1}\left[\begin{pmatrix} \beta^{n-1} \\ 0 \end{pmatrix} + c_n \begin{pmatrix} 0 \\ \mathbf{J}\bar{\beta}^{n-1} \end{pmatrix}\right] = \begin{pmatrix} P_n + c_n\bar{\mathbf{z}}'\bar{\beta}^{n-1} \\ 0 \\ \vdots \\ 0 \\ c_n P_n + \mathbf{z}'\beta^{n-1} \end{pmatrix}.$$

To eliminate the last term, choose $c_n = -\mathbf{z}'\beta^{n-1}/P_n$. This is called the

1. Background Material

reflection coefficient. Now

$$\alpha^n = \begin{pmatrix} \beta^{n-1} \\ 0 \end{pmatrix} + c_n \begin{pmatrix} 0 \\ J\bar{\beta}^{n-1} \end{pmatrix}$$

since we assumed $\det(\mathbf{R}_n) \neq 0$.

Theorem 5.1 $P_{n+1} = P_n(1 - |c_n|^2)$.
This is an algorithm for predicting the output y and its covariance, and it is better than sequential filtering if only the output predictions are needed and the process is stationary.

Remark: If the c_n sequence is known, α_i^n can be found for all n, i.
Burg approximated c_n by

$$\hat{c}_n = -\frac{2\sum(F_i B_i)}{\sum F_i^2 + \sum B_i^2},$$

where F_i is the forward error, and B_i is the backward error, $\tilde{y}_0$, given $y_1, ..., y_i$.
Note: $|\hat{c}_n|^2 < 1$, unless the process is deterministic, which we assume is not the case. For the first c_i: $E(y_2 - \alpha y_1)^2 = r_0 - \alpha(r_1 + r_{-1}) + \alpha^2 r_0$. To minimize this, $\alpha = (r_1 + r_{-1})/2r_0$. These reflection coefficients are motivated by geophysical applications (detecting underground formations by sound wave reflection); see Clairbout [22]. Burg used these reflection coefficients to compress and transmit speech.

$$\tilde{y}_n = \sum_{i=0}^{n} \beta_i^{n-1} y_i = u_n$$

or white noise. So this is an autoregressive process. One can approximate the spectrum of the discrete process with

$$\frac{1}{\beta(s)\beta(\frac{1}{s})},$$

where $\beta(s)$ denotes the polynomial whose coefficients are the β_i^{n-1}; i.e., the reflection coefficients give the "spectrum" of the output for a stationary process.

A fact leading to the widespread use of the Burg technique is that the $\hat{c}_n$s produce a stable process. The β_i^{n-1}s are useful for identifying the observation process, but the c_ns are statistically too irregular to be of much use for characterizing the process. For details on identification of autoregressive moving average processes using the Burg technique, see [23].

Now, for $d \neq 1$ (no time-varying system coefficients), $\mathbf{x}_{n+1} = \mathbf{\Phi}\mathbf{x}_n + \mathbf{G}\mathbf{u}_n$, $\mathbf{x}_0 = \mathbf{c}$, $\mathbf{c} \in N(0, \Gamma)$, $E\mathbf{u}_n\mathbf{u}'_n = \mathbf{I}\delta_{n,m}$, and $\mathbf{y}_n = \mathbf{H}\mathbf{x}_n$, what happens as $t \to \infty$? The quantity

$$\mathbf{S}_{n+1} = E\mathbf{x}_{n+1}\mathbf{x}'_{n+1}$$

will in general not be independent of n, unless a proper choice of Γ is made. Assume $\det(s\mathbf{I} - \mathbf{\Phi})$ has roots inside the unit disk, so that as $n \to \infty$, $\mathbf{x}_n \to$ constant.

$$\mathbf{S}_{n+1} = E(\mathbf{\Phi}\mathbf{x}_n + \mathbf{G}\mathbf{u}_n)(\mathbf{\Phi}\mathbf{x}_n + \mathbf{G}\mathbf{u}_n)' = \mathbf{\Phi}\mathbf{S}_n\mathbf{\Phi}' + \mathbf{G}\mathbf{G}'.$$

Suppose that

$$\mathbf{S}_n = \mathbf{\Phi}^n\Gamma\mathbf{\Phi}'^n + \sum_{i=0}^{n-1}\mathbf{\Phi}^i\mathbf{G}\mathbf{G}'\mathbf{\Phi}'^i$$

is a solution. Since roots are ≤ 1, $\mathbf{\Phi}^n$ converges as $n \to \infty$, and therefore

$$\mathbf{S}_n \to \mathbf{S}_\infty = \sum_{i=0}^{\infty}\mathbf{\Phi}^i\mathbf{G}\mathbf{G}'\mathbf{\Phi}'^i.$$

So the $\mathbf{x}_n, \mathbf{y}_n$ will be stationary, i.e., $E\mathbf{x}_n\mathbf{x}'_m = F(n-m)$. A fixed point of this equation is $\mathbf{S} = \mathbf{\Phi}\mathbf{S}\mathbf{\Phi}' + \mathbf{G}\mathbf{G}'$.

Using $\otimes$, the Kronecker or tensor product (see [24]) and its property that the eigenvalues of the Kronecker product are the product of the eigenvalues

1. Background Material

of the constituent matrices,

$$(\mathbf{I} - \boldsymbol{\Phi} \otimes \boldsymbol{\Phi})\mathbf{s} = \mathbf{g}, \quad \text{where } \mathbf{g} \text{ is an } n^2 \text{ vector.} \tag{5.3}$$

To get $\mathbf{s}$ and $\mathbf{g}$, stack the columns of $\mathbf{S}, \mathbf{G}$. Then the eigenvalues of $(\mathbf{I} - \boldsymbol{\Phi} \otimes \boldsymbol{\Phi})$ are $1 - \lambda_i \lambda_j > 0$ so that Equation (5.3) has a solution.

Theorem 5.2 *If $(\boldsymbol{\Phi}, \mathbf{G})$ is completely controlable and $\boldsymbol{\Phi}$ is strictly stable then $\mathbf{S} > 0$.*

(Try proving this as an exercise.)

Let $\boldsymbol{\Gamma} = \mathbf{S}$; i.e., the starting covariance is the stationary distribution

$$E(\mathbf{H}\mathbf{x}_n \mathbf{x}'_m \mathbf{H}') = \mathbf{H} E(\mathbf{x}_n \mathbf{x}'_m) \mathbf{H}' \quad \text{by linearity}$$

$$= \begin{cases} \mathbf{H}\boldsymbol{\Phi}^{n-m}\mathbf{S}\mathbf{H}', & \text{if } n > m; \\ \mathbf{H}\mathbf{S}\boldsymbol{\Phi}'^{m-n}\mathbf{H}', & \text{if } n < m. \end{cases}$$

Burg is actually computing the innovations, $\mathbf{y_n} - \hat{\mathbf{y}}_\mathbf{m}$. We can do this in the multidimensional case:

$$\mathbf{T} \begin{pmatrix} R_0 & R_1 & \cdots \\ R'_1 & R_0 & \cdots \\ \vdots & \vdots & \ddots \end{pmatrix} \mathbf{T}' = \begin{pmatrix} I_1 & & \\ & \ddots & \\ & & I_n \end{pmatrix}.$$

So

$$\begin{pmatrix} I & 0 & 0 & \cdots \\ A & I & 0 & \cdots \\ \vdots & \vdots & \ddots & \\ \vdots & \vdots & \vdots & \ddots \end{pmatrix} \mathbf{y}^n = \begin{pmatrix} I_0 \\ I_1 \\ \vdots \\ I_n \end{pmatrix}$$

or $\mathbf{L}\mathbf{y}^n = \mathbf{I}^n$. The covariance of this is

$$\mathbf{L}\mathbf{R}\mathbf{L}' = \begin{pmatrix} P_0 & 0 & \cdots \\ 0 & \ddots & 0 \\ 0 & 0 & P_n \end{pmatrix},$$

which is an $s \times s$ matrix. Recall the notation $c^n = (c_1, ..., c_n)$.

The model we had here was not autoregressive. $\beta(D)y = u$ is autoregressive, and $y = \beta(D)u$ is a moving average. $\beta(D)y = \gamma(D)u$ is autoregressive moving average, where β and γ are polynomials and D is the unit delay operator.

Burg is in a sense estimating the covariance by computing the c_ns.

Lecture 6

Signal Processing

1 The Burg Technique

From the last lecture, Burg defined a sequential method to determine "optimal" forward and backward predictors from a given random sequence. These equations take the form

$$f_N(z) = f_{N-1}(z) + \hat{c}_N z b_{N-1}(z)$$

$$b_N(z) = z b_{N-1}(z) + \overline{\hat{c}}_N f_N(z).$$

Here, z is the classical z transform variable, the unit delay operator. The $\hat{c}$ are *estimated* reflection coefficients, so that the Burg predictor is really only an approximation of the optimal predictor. For applications of the Burg predictor to speech pattern recognition, see reference [25].

Assuming a waveform moving through a refractive medium, an analogy between the Burg reflection coefficients and the reflected wave can be made. In the following equation, the first term represents the incident wave, which is equal to the sum of the transmitted wave (the first term to the right of the equal sign) and the reflected wave (the second term to the right of the equal sign). The forward equation is

$$\begin{pmatrix} 1 \\ a_1^N \\ \vdots \\ a_{N-1}^N \\ a_N^N \end{pmatrix} = \begin{pmatrix} 1 \\ a_1^{N-1} \\ \vdots \\ a_{N-1}^{N-1} \\ 0 \end{pmatrix} + c_N \begin{pmatrix} 0 \\ \bar{a}_{N-1}^{N-1} \\ \vdots \\ \bar{a}_1^{N-1} \\ 1 \end{pmatrix}, \qquad (6.1)$$

and the backward equation can be found by applying the reversal operator, J, to the a vector, which results in the following:

$$\begin{pmatrix} \bar{a}_N^N \\ \bar{a}_{N-1}^N \\ \vdots \\ \bar{a}_1^N \\ 1 \end{pmatrix} = \begin{pmatrix} 0 \\ \bar{a}_{N-1}^{N-1} \\ \vdots \\ \bar{a}_1^{N-1} \\ 1 \end{pmatrix} + \bar{c}_N \begin{pmatrix} 0 \\ a_{N-1}^{N-1} \\ \vdots \\ a_1^{N-1} \\ 1 \end{pmatrix}. \qquad (6.2)$$

Take the dot of Equation (6.1) with a vector of unit delays on the left to obtain

$$\left(z^0, z^1, \ldots, z^N\right) \begin{pmatrix} 1 \\ a_1^N \\ \vdots \\ a_N^N \end{pmatrix} = f_N(z)$$

(where zs are unit delay); this is the forward predictor error filter.

The z-transformation of the N-stage wave front moving through the refractive medium is obtained by the z-transformation of N-stage incident wave, the z-transformation of N-stage reflected wave and N-stage z-transformation of the refracted wave. Taking the dot product of the right-hand side of Equation (6.1) yields

$$f_N(z) = f_{N-1}(z) + c_N \, z \, b_{N-1}(z)$$

and hence by analogous operations on Equation (6.2), the backward error filter is given by

$$b_N(z) = z b_{N-1} + \bar{c}_N f_N(z).$$

1. The Burg Technique

The prediction error for the random sequence is given by

$$\varepsilon_N = \tilde{y}_N + a_1^N \tilde{y}_{N-1} + \ldots + a_N^N \tilde{y}_0.$$

It follows that the a_N sequences introduced in Chapter 5 are now completely explicitly specified by the knowledge of c_N in the (recursive) solution of the following equation:

$$R_N \begin{pmatrix} 1 \\ a_1^{N-1} \\ \vdots \\ a_{N-1}^{N-1} \end{pmatrix} = \begin{pmatrix} P_{N-1} \\ 0 \\ \vdots \\ 0 \end{pmatrix}.$$

Example 6.1 *The following 2×2 example illustrates the above procedure:*

$$\begin{pmatrix} r_0 & r_{-1} \\ r_1 & r_0 \end{pmatrix} \begin{pmatrix} 1 \\ \alpha_1^1 \end{pmatrix} = \begin{pmatrix} P_1 \\ 0 \end{pmatrix}$$

$$\Rightarrow \quad \alpha_1^1 = -\frac{r_1}{r_0} \stackrel{\triangle}{=} c_1$$

since

$$\begin{pmatrix} 1 \\ \alpha_1^1 \end{pmatrix} = \begin{pmatrix} 1 \\ 0 \end{pmatrix} + c_1 \begin{pmatrix} 0 \\ 1 \end{pmatrix},$$

and hence

$$\Rightarrow \quad P_1 = r_0(1 - |c_1|^2), \quad \text{since } r_{-1} = \bar{r}_1.$$

In a sense then, $R_i \to f_N, b_N, c_N$ and likewise $c_N \to f_N, b_N$. This is called Levinson recursion. In practice, the order N can be large for much data.

Given an autoregressive moving average (ARMA) sequence

$$k(z)y = m(z)u,$$

where $k(z)$ and $m(z)$ are polynomials, estimating m is difficult [26]. Burg estimates only autoregressive processes of the form

$$k(z)y = u.$$

The moving average in ARMA refers to "numerator dynamics" [26]. If we let the parameter m denote the number of independent looks at the same thing, then the forward error, f_m, is the error for the mth look at the sequence of xs and is given as

$$f_m = a_{N-1}\tilde{y}_{2,m} + \ldots + a_1\tilde{y}_{n,m} + \tilde{y}_{N+1,m},$$

and the backward error is

$$b_m = \tilde{y}_{1,m} + \bar{a}_1\tilde{y}_{2,m} + \ldots + \bar{a}_{N-1}\tilde{y}_{N,m}.$$

Burg had the revolutionary idea of approximating the value of the reflection coefficient by minimizing the RSS of the forward and backward error, which results in the following expression:

$$\hat{c}_N = -\frac{2\sum \bar{b}_m f_m}{\sum b_m \bar{b}_m + \sum f_m \bar{f}_m}.$$

2 Signal Processing

Burg's update equation for the forward prediction error filter uses the forward/backward iteration to find the reflection coefficients:

$$f_N(z) = f_{N-1}(z) + c_N z b_{N-1}(z)$$

$$b_N(z) = z b_{N-1} + \bar{c}_N f_N(z).$$

2. Signal Processing

Example 6.2 *The following illustrates the iteration technique of Burg-1:*

$$f_1 = \begin{pmatrix} 1 \\ c_1 \end{pmatrix} \quad b_1 = \begin{pmatrix} c_1 \\ 1 \end{pmatrix} \quad f_1 = 1 + c_1 z$$

$$f_2 = \begin{pmatrix} 1 \\ c_1 \\ 0 \end{pmatrix} + c_2 \begin{pmatrix} 0 \\ c_1 \\ 1 \end{pmatrix}$$

$$\rightarrow \quad f_2(z) = 1 + c_1(1+c_2)z + c_2 z$$

with

$$b_2 = \begin{pmatrix} c_2 \\ c_1(1+c_2) \\ 0 \end{pmatrix}$$

$$f_3 = \begin{pmatrix} 1 \\ c_1(1+c_2) \\ c_2 \\ 0 \end{pmatrix} + c_3 \begin{pmatrix} 0 \\ c_2 \\ c_1(1+c_2) \\ 1 \end{pmatrix}$$

$$f_3(z) = 1 + (c_1(1+c_2) + c_3 c_2)z + (c_2 + c_3 c_2))z^2 + c_3 z^3.$$

The following program, written in BASIC, shows an implementation of the Burg-1 recursive algorithm:

```
0 REM   BURG1
1 P= 0
5 INPUT "# OF REFLECTION COEFFICIENTS";M
10 DIM F(1000),C(M),A(M+ 1,M+ 1),FF(1000)
12 DIM B(1000),S(M)
15 S0= 0:S1= 0
16 OPEN "F.ARY" FOR INPUT AS #1
17 INPUT #1,F(0):N= F(0):FOR J=1 TO N:INPUT #1, F(J)
18 NEXT J
20 FOR I= 1 TO N
30 S0= S0+ F(I)^ 2/ N
32 S1= S1+ F(I)/ N
35 NEXT
40 FOR I= 1 TO M
50 A(I,P)= 0
```

```
60 NEXT
70 A(M+ 1,P)= SQR (S0- S1^ 2)
75 S(0)= S0- S1^ 2
80 FOR I= 1 TO N
90 B(I)= F(I)
95 FF(I)= F(I)
100 NEXT
110 P= P+ 1
120 S1= 0
130 S2= 0
140 S3= 0
150 FOR I= 1 TO N
160 IF I< P+ 1 THEN 180
170 S1= S1+ FF(I)* B(I- P)
175 S2= S2+ B(I)* B(I)
180 IF I> N- P GOTO 200
190 S3= S3+ FF(I)* FF(I)
200 NEXT
210 A(P,P)= 2* S1/ (S2+ S3)
211 PM= P- 1
212 IF  PM< 1 THEN  220
213 FOR  I= 1 TO  PM
214 A(I,P)= A(I,PM)- A(P,P)* A(P- I,PM)
215 NEXT
220 S(P)= (1- A(P,P)* A(P,P))* S(P- 1)
225 PRINT P,A(P,P)
227 PRINT "SPECTRAL COEFFICIENTS"
228 FOR I= 1 TO P:PRINT A(I,P):NEXT
230 P1= P+ 1
240 FOR  I= P1 TO  M
250 A(I,P)= 0
260 NEXT
270 A(M+ 1,P)= SQR (S(P))
280 FOR  I= P1 TO  N
290 S1= 0
300 FOR  K= 1 TO  P
310 S1= S1+ A(K,P)* F(I- K)
320 NEXT
330 FF(I)= F(I)- S1
340 NEXT
350 NM= N- P
360 FOR  T= 1 TO  NM
370 S1= 0
380 FOR  K= 1 TO  P
390 S1= S1+ A(K,P)* F(T+ K)
400 NEXT
410 B(T)= F(T)- S1
420 NEXT
430 IF  P= M THEN  500
440 GOTO  110
500 END
```

Hence, we can describe a random process by the reflection coefficients to cut down on the bandwidth necessary for transmission. An example of this is the transmission of human speech by representing it as 9 to 11 reflection coefficients over small intervals of time. Most cognitive speech processing uses 13 coefficients to define a pattern.

3 Burg Revisited (Rouché's Theorem)

Burg's update equation for the prediction filter is

$$f_{n+1}(z) = f_n(z) + \hat{c}_n z b_n(z), \quad |\hat{c}_n| < 1.$$

Burg shows that f_{n+1}, the prediction error filter, has the same number of zeros inside or on the unit circle as f_n by using Rouché's theorem. The proof of the asymptotic stability of the prediction filter is an induction proof for $n = 1$, $1 + \hat{c}_n z$ has zeros outside the unit disk and assuming that the zeros of $f_n(z)$ are outside the unit disk, the induction step follows from the following theorem, with $f(z) = f_n(z)$ and $g(z) = \hat{c}_n z b_n(z)$.

Theorem 6.1 (Rouché) *If f,g are analytic in D and δD (the boundary of D), and $|f| > |g|$ on δD. Then f and $f+g$ have the same number of zeros, counting multiplicities, in D.*

Consider the following contour integral:

$$\frac{1}{2\pi i} \ln f \Big|_{\delta C} = \frac{1}{2\pi i} \oint_C \frac{f'}{f} dz.$$

As we integrate the contour around a complete revolution, we flip up onto another sheet of the Riemann surface at the cut. Assume that there are no poles or zeros on the boundary. If there are zeros at ξ_i of multiplicity r_i within the contour and if there are poles at ρ_i of multiplicity s_i, then the

Laurent expansion of f is given as

$$f(z) = \frac{\prod_{i=1}^{k}(z-\xi_i)^{r_i}}{\prod_{j=1}^{\ell}(z-\rho_j)^{s_j}} g(z),$$

where $g(z)$ has neither poles nor zeros within or on C. Then,

$$f' = f\left(\sum_{i=1}^{k}\frac{r_i}{z-\xi_i} - \sum_{j=1}^{\ell}\frac{s_j}{z-\rho_j}\right)$$

$$\Rightarrow \quad \frac{f'}{f} = \sum_{i=1}^{k}\frac{r_i}{z-\xi_i} - \sum_{j=1}^{\ell}\frac{s_j}{z-\rho_j}$$

$$\Rightarrow \quad \oint_C \frac{f'}{f} dz = 2\pi i \left(\sum r_i - \sum s_i\right)$$

$$= 2\pi i (\text{number of zeros} - \text{number of poles}).$$

The following is Caratheodory's proof of Rouché's theorem: Assume the compact set D and its boundary δD and the two functions, f and g, defined on D such that

$$0 < m = \inf_{z \in \delta D}\{|f| - |g|\}.$$

As an aside, note for $0 \le \lambda \le 1$ the sum

$$|f + \lambda g| \ge \Big||f| - \lambda|g|\Big| \quad \Rightarrow \quad |f| - \lambda|g| \ge m.$$

Hence,

$$\frac{1}{2\pi i}\int_{\delta D}\frac{f' + \lambda g'}{f + \lambda g}dz = \text{integer}$$

$$= k(\lambda)$$

since we have a continuous and bounded denominator; $k(\lambda)$ is continuous. Hence,

$$k(0) = k(1),$$

which implies that the number of zeros of f is the same as the number of zeros of $f + g$.

Contrast this with a more standard proof:

$$\frac{f' + g'}{f + g} = \frac{f'}{f} + \frac{g'/f - f'g/f^2}{(1 + g/f)}$$

$$= \frac{f'}{f} + \frac{d(g/f)}{(1 + g/f)}.$$

Integrate

$$\int_{\delta D} \frac{f' + g'}{f + g} dz = 2\pi i N + \int_{\delta D} \frac{d(g/f)}{(1 + g/f)}.$$

However, this implies that the integration around the boundary that does not enclose the origin is zero, and we are finished. Here we see that Caratheodory's proof is much nicer.

Therefore, f_{n+1} and f_n both have no zeros inside or on the unit circle, and hence the asymptotic stability is demonstrated for the case where c_i are less than one for all i.

3.1 Burg's Inverse Iteration

The Burg inverse iteration scheme (called Burg-2 here) was developed in reference [27]. This scheme was Burg's attempt to put a metric on the covariance function and then to try to minimize the metric (and hence the covariance) with the proper estimate. This approach overcomes two of the shortcomings of the Burg-1 technique described previously since it

1. gets around the fact that the reflection coefficients of Burg-1 are fixed for previous data, and

2. results in covariance estimates that are based on the whole sequence of observations.

Assume the complex domain for the following derivation. Define the norm:

$$||\mathbf{z}||_A^2 = \mathbf{z}^* A \mathbf{z}, \quad A \text{ hermitian},$$

where $(*)$ denotes the complex conjugate transpose.

Assume a spherical Gaussian distribution, if $\mathbf{z} \in C^n$,

$$P(\mathbf{z}) = [(\pi)^n (\det A)]^{-1} \exp(-||\mathbf{z} - \mu||_{A^{-1}}^2).$$

The complex case can be simpler (e.g., a chi-squared distribution for complex random variables becomes the beta distribution). This can be seen as the characteristic function of chi-squared and beta:

$$\chi(i\omega) = (1 + i\omega)^{-n/2} \quad \text{are equivalent to} \quad \beta(i\omega) = (1 + i\omega)^{-n}$$

respectively.

Assume observations $\mathbf{z}_1, \ldots, \mathbf{z}_n$, are independent, identically distributed random variables with variance $R = E\{\mathbf{z}_i \mathbf{z}_i^*\}$. The probability distribution function for the Nth iterate of the $\mathbf{z}^n$ in the complex case is given by

$$P(z_1^n \cdots z_N^n) = [(\pi)^{Nn}(\det R)^N]^{-1} \exp\left(-\sum_{i=1}^N ||\mathbf{z}_i^n||_{R^{-1}}^2\right).$$

Develop the maximum-likelihood estimator as follows:

$$-\ln P = L \stackrel{\Delta}{=} N \ln(\det R) + N \operatorname{trace}\{R^{-1} S\},$$

where the sample covariance, S, is given by

$$S = 1/N \sum_{i=1}^N \mathbf{z}_i \mathbf{z}_i^*.$$

3. Burg Revisited (Rouché's Theorem)

Note that since R is Hermitian,

$$\ln\{\det R\} = \text{trace} \ln R$$

$$\ln\{\lambda_1 \lambda_2 \cdots \lambda_n\} = \sum_{i=1}^{n} \ln \lambda_i;$$

hence, the likelihood ratio is

$$\frac{L}{N} = \text{trace}\{\ln R + R^{-1}S\}.$$

Now find the maximum-likelihood estimate of S assuming that R is expandable on a basis of Q_is; i.e., $R = \sum_{i=1}^{n} \alpha_i Q_i$; take the first variation and let it vanish to obtain

$$\frac{\delta L}{N} = \text{trace}\{R^{-1}\delta R - R^{-1}\delta R R^{-1}S\},$$

and since $\text{trace}\{AB\} = \text{trace}\{BA\}$, then $\delta L = 0 \Rightarrow$

$$\text{trace}\{R^{-1} - R^{-1}SR^{-1}\}\delta R = 0.$$

Hence, inverse iteration is defined as finding S_1 so that the variational equation is satisfied; i.e.,

$$\text{trace}\{R^{-1}RR^{-1} - R^{-1}S_1 R^{-1}\}\delta R = 0$$

or find δS so that

$$\text{trace}\{R^{-1}RR^{-1} - R^{-1}(S + \delta S)R^{-1}\}\delta R = 0.$$

Now, the new R is $R_1 = R - \delta S$, so that

$$\text{trace}\{R^{-1}R_1 R^{-1} - R^{-1}SR^{-1}\}\delta R = 0.$$

This gives a linear equation to solve for R_1. Assuming that

$$R_1 = \sum_{i=1}^{n} \alpha_i Q_i$$

the variational equation simplifies to the form $A\mathbf{x} = \mathbf{b}$, where

$$A = \{a_{ij}\}$$

$$a_{ij} = \text{trace}\{R^{-1}Q_i R^{-1}Q_j\}$$

$$b_i = \text{trace}\{R^{-1}SR^{-1}Q_i\}$$

$$\mathbf{x} = (\alpha_1, \alpha_2, \ldots, \alpha_n)'.$$

Example 6.3 *Suppose the process is real and stationary; then R is Toeplitz. It is easy to see that the 3×3 Toeplitz matrices have the following basis:*

$$Q_1 = \begin{bmatrix} 1 & 0 & 0 \\ 0 & 1 & 0 \\ 0 & 0 & 1 \end{bmatrix} \tag{6.3}$$

$$Q_2 = \begin{bmatrix} 0 & 1 & 0 \\ 1 & 0 & 1 \\ 0 & 1 & 0 \end{bmatrix} \quad \tilde{Q}_2 = \begin{bmatrix} 0 & i & 0 \\ -i & 0 & i \\ 0 & -i & 0 \end{bmatrix}$$

$$Q_3 = \begin{bmatrix} 0 & 0 & 1 \\ 0 & 0 & 0 \\ 1 & 0 & 0 \end{bmatrix} \quad \tilde{Q}_3 = \begin{bmatrix} 0 & 0 & i \\ 0 & 0 & 0 \\ -i & 0 & 0 \end{bmatrix}.$$

$\vdots$

Example 6.4 *Given a 2 by 2 system with*

$$S = \begin{pmatrix} a & z \\ \bar{z} & b \end{pmatrix}, \quad \text{find } R = \begin{pmatrix} \alpha & \beta \\ \bar{\beta} & \alpha \end{pmatrix}, \quad \alpha \text{ real.}$$

3. Burg Revisited (Rouché's Theorem)

This method is an alternative to Burg-1. It allows fitting to all data in a single batch as contrasted to the previous reflection coefficient method where each coefficient was fixed as each measurement was processed sequentially.

The following program, written in BASIC, shows an implementation of the Burg-2 algorithm, for complex 3×3 matrices:

```
5 REM   BURG2
10 DIM A(3,3,11),C(3,3),I(3,3),J(3,3),R(3,3),R1(3,3),L(3),X(3)
15 DIM S(3,3),T(3,3),SM(6),SS(11)
20 OPEN "SM.ARY" FOR INPUT AS #1:FOR J=1 TO 6
21 INPUT #1,SM(J):NEXT J
22 OPEN "Ss.ARY" FOR INPUT AS #2:FOR J=1 TO 11
23 INPUT #2,SS(J):NEXT J
24 A1= SS(1):A0= SS(2)
25 S(1,1)= SM(1):S(1,2)= SM(2):S(1,3)= SM(3):S(2,2)= SM(4)
26 S(2,3)= SM(5):S(3,3)= SM(6)
30 S(2,1)= S(1,2):S(3,1)= S(1,3):S(3,2)= S(2,3)
31 PRINT SS(6),SS(7),A1* SS(7)+ A0* SS(6)
32 PRINT "THIS IS TRUE COVARIANCE "
33 PRINT SS(7),SS(6),SS(7)
34 PRINT   A1* SS(7)+ A0* SS(6),SS(7),SS(6)
36 DD= SS(6)^ 3+ 2* SS(7)^ 2* SS(8)- SS(8)^ 2* SS(6)
     - 2* SS(6)*SS(7)^ 2
38 PRINT "DETERMINANT OF TRUE=",DD
40 FOR I= 1 TO 3:FOR J= 1 TO 3
45 A(I,J,6)= S(I,J):R(I,J)= 0:A(I,J,7)= 0
46 A(I,J,8)= 0:A(I,J,9)=0
50 IF I= J THEN R(I,J)= 1:A(I,J,7)= 1
60 NEXT :NEXT
70 A(1,3,9)= 1:A(3,1,9)= 1
75 A(1,2,8)= 1:A(2,1,8)= 1:A(3,2,8)= 1:A(2,3,8)= 1
80 V= S(1,1)+ S(2,2)+ S(3,3)
90 FOR K9= 1 TO 20
100 FOR I= 1 TO 3:FOR J= 1 TO 3
110 I(I,J)= R(I,J)
120 NEXT :NEXT
130 GOSUB 9400
135 FOR I= 1 TO 3:FOR J= 1 TO 3
140 A(I,J,5)= J(I,J)
145 NEXT :NEXT
150 PRINT "DETERMINANT=",D
160 FOR   I= 1 TO 3:FOR   J= 1 TO   3
170 PRINT A(I,J,5);:NEXT
180 PRINT :NEXT
185 REM
190 PRINT "PREDICTOR",A(1,2,5)/ A(1,1,5),A(1,3,5)/ A(1,1,5)
200 PRINT "INTERPOLATOR",A(1,2,5)/ A(2,2,5),A(3,2,5)/ A(2,2,5)
230 REM
```

```
310 I1= 5:I2= 6
320 GOSUB 9100
330 J1= 4:GOSUB 9200
340 V= LOG (D)+ A(1,1,4)+ A(2,2,4)+ A(3,3,4)
350 I2= 7:GOSUB 9100
360 J1= 1:GOSUB 9200
370 I2= 8:GOSUB 9100
380 J1= 2:GOSUB 9200
390 I2= 9:GOSUB 9100
400 J1= 3:GOSUB 9200
410 REM   COEF.OF LINEAR EQ
415 I1= 1
420 I2= 1:GOSUB 9100
430 GOSUB 9300
440 T(1,1)= T
450 I2= 2:GOSUB 9100
460 GOSUB 9300
470 T(1,2)= T:T(2,1)= T
490 I2= 3:GOSUB 9100
500 GOSUB 9300
510 T(1,3)= T:T(3,1)= T
520 I1= 2:GOSUB 9100
530 GOSUB 9300
540 T(2,3)= T:T(3,2)= T
550 I1= 3:GOSUB 9100
560 GOSUB 9300
570 T(3,3)= T
580 I1= 2:I2= 2:GOSUB 9100
590 GOSUB 9300
600 T(2,2)= T
610 I1= 4:I2= 1:GOSUB 9100
620 GOSUB 9300
630 L(1)= T
640 I2= 2:GOSUB 9100
650 GOSUB 9300
660 L(2)= T
670 I2= 3:GOSUB 9100
680 GOSUB 9300
690 L(3)= T
700 FOR I= 1 TO 3:FOR J= 1 TO 3
710 I(I,J)= T(I,J)
720 NEXT :NEXT
730 GOSUB 9400
740 FOR M= 1 TO 3
750 X(M)= 0
760 FOR K= 1 TO 3
770 X(M)= X(M)+ J(M,K)* L(K)
780 NEXT :NEXT
800 FOR I= 1 TO 3
810 R1(I,I)= X(1)
820 IF I< 3 THEN R1(I+ 1,I)= X(2):R1(I,I+ 1)= X(2)
```

3. Burg Revisited (Rouché's Theorem)

```
830 NEXT
840 R1(1,3)= X(3):R1(3,1)= X(3)
850 IF R1(1,1)< 0 THEN 7000
860 IF R1(1,1)* R1(2,2)- R1(1,2)^ 2< 0 THEN 7000
870 FOR I= 1 TO 3:FOR J= 1 TO 3
880 I(I,J)= R1(I,J)
890 NEXT :NEXT
900 GOSUB 9400
910 IF D< 0 THEN 7000
920 FOR I= 1 TO 3:FOR J= 1 TO 3
930 A(I,J,0)= J(I,J)
940 NEXT :NEXT
950 I1= 0:I2= 6:GOSUB 9100
960 GOSUB 9300
970 V1= T+ LOG (D)
980 IF V1>V THEN 7000
985 V= V1
990 FOR I= 1 TO 3:FOR J= 1 TO 3
1000 R(I,J)= R1(I,J)
1010 NEXT :NEXT
1015 PRINT K9:FOR I= 1 TO 3:FOR J= 1 TO 3
1017 PRINT R(I,J);:NEXT
1018 PRINT
1019 NEXT
1020 NEXT
1030 END
7000 REM   JUMP OUT INADMISSIBLE R
7010 FOR I= 1 TO 3:FOR J= 1 TO 3
7020 A(I,J,10)= R(I,J)+ 1/ 2* (R1(I,J)- R(I,J))
7025 R1(I,J)= A(I,J,10)
7030 NEXT :NEXT
7040 GOTO 850
9100 REM   SUB MUL
9110 FOR I= 1 TO 3:FOR J= 1 TO 3
9120 C(I,J)= 0
9130 FOR K= 1 TO 3
9140 C(I,J)= C(I,J)+ A(I,K,I1)* A(K,J,I2)
9150 NEXT
9160 NEXT :NEXT
9170 RETURN
9200 REM   UNLOAD
9210 FOR I= 1 TO 3:FOR J= 1 TO 3
9220 A(I,J,J1)= C(I,J)
9230 NEXT :NEXT
9240 RETURN
9300 REM   TRACE
9310 T= 0
9320 FOR I= 1 TO 3
9330 T= T+ C(I,I)
9340 NEXT
9350 RETURN
```

```
9400 REM    INVERSE
9410 D= I(1,1)* (I(2,2)* I(3,3)- I(2,3)^ 2)- I(1,2)*
          (I(1,2)* I(3,3)- I(3,1)* I(2,3))
9420 D= D+ I(1,3)* (I(1,2)* I(3,2)- I(2,2)* I(3,1))
9430 J(1,1)= 1/ D* (I(2,2)* I(3,3)- I(2,3)^ 2)
9440 J(1,2)= - 1/ D* (I(1,2)* I(3,3)- I(3,1)* I(2,3))
9450 J(2,1)= J(1,2)
9460 J(1,3)= 1/ D* (I(1,2)* I(3,2)- I(2,2)* I(3,1))
9470 J(3,1)= J(1,3)
9480 J(2,2)= 1/ D* (I(1,1)* I(3,3)- I(1,3)^ 2)
9490 J(3,3)= 1/ D* (I(1,1)* I(2,2)- I(1,2)^ 2)
9500 J(2,3)= - 1/ D* (I(1,1)* I(3,2)- I(1,2)* I(1,3))
9510 J(3,2)= J(2,3)
9520 RETURN
9530 END
```

Lecture 7

Classical Approach

1 Classical Steady-State Filtering

The following will illustrate the classical Wiener approach to a discrete filtering problem followed by the Riccati equation approach in a later chapter. The development here follows [29]; note that in all that follows, t takes on integer values, and $t_0 = -\infty$:

$$x_{n+1} = \lambda x_n + u_n, \quad |\lambda| < 1$$

$$z_n = x_n + v_n, \quad E v_i v_j = \delta_{ij} r.$$

Define the following for the scalar stationary process, $z(t)$:

$$E\left\{z(t+\tau)z'(t)\right\} = o(\tau) = m\lambda^{|\tau|} + r$$

$$E\left\{x(t+\tau)z'(t)\right\} = c(\tau) = m\lambda^{|\tau|}$$

$$E\left\{x(t+\tau)x'(t)\right\} = s(\tau) = m\lambda^{|\tau|}$$

where $m = \frac{1}{1-\lambda^2}$ for all t, τ since the process is stationary. Recall we are going to look at the steady-state estimate:

$$\hat{x}(t_2|t_1) = \sum_{j=0}^{\infty} k_j z(t_1 - j).$$

Now, Wiener-Hopf says the error is perpendicular to the data; i.e.,

$$E\{\hat{x}(t_2|t_1)z'(s)\} = E\{x(t_2)z'(s)\}, \quad \forall \; -\infty < s \leq t_1,$$

which implies

$$c(t_2 - s) = \sum_{j=0}^{\infty} k_j \, o(t_1 - j - s), \quad \forall \; s \leq t_1.$$

Consider the case $t_1 = t_2$ (as described in [29]):

$$c(\lambda) = \sum_{\mu=0}^{\infty} k_\mu \, o(\lambda - \mu), \quad \lambda \geq 0.$$

In addition, note that covariances can be factored since their spectral densities are of the form $f(s) = f(\frac{1}{s})$; i.e., they are invariant under the mapping $s \to \frac{1}{s}$. Hence since o is a covariance, write

$$o(\lambda) = \sum_{i=-\infty}^{\infty} a_{\lambda-i} \bar{a}_i$$

with conditions

1. $a_i = 0$, $i < 0$, $\sum_{i=0}^{\infty} |a_i| < \infty$,

2. $\bar{a}_i \equiv a_{-i}$,

3. Laplace transform, $a(s) = \sum_{i=0}^{\infty} a_i s^i$, has no zeros in the set $|s| \leq 1$.

This is the analog of multiplicative spectral factorization. It is a Banach algebra (i.e., a complete, normed linear space) since

$$\| a * b \| \leq \| a \| \| b \|.$$

Note that condition number three results from requiring that a_i has an inverse. As an illustration of this condition, say that sequence a_{i-j} is

1. Classical Steady-State Filtering

multiplied by sequence b_j:

$$\sum_{j=-\infty}^{\infty} a_{i-j} b_j = \delta_{i,0}.$$

Multiply both sides by s^i and sum over all i:

$$\sum_i s^i \sum_{j=0}^{i} a_{i-j} b_j = \sum_{i=0}^{\infty} \delta_{i,0} s^i = 1.$$

Since a, b vanish for negative indices:

$$\sum_i s^{i-j} \sum_j^{i} a_{i-j} s^j b_j.$$

Now interchange sums to obtain the discrete Laplace transform:

$$a(s)b(s) = 1$$

$$b(s) = \frac{1}{a(s)}.$$

Hence, we cannot have zeros in $a(s)$ or we cannot perform the above inversion.

Example 7.1 *Given the state covariance, $\mathcal{S}$, find the factorization as follows:*

$$\mathcal{S}(\tau) = \frac{\beta^\tau}{1-\beta^2} \quad |\beta| < 1.$$

Factor the right-hand side

$$\frac{\beta^j}{1-\beta^2} = \sum a_{j-k} \bar{a}_k$$

$$= \sum_{k=-\infty}^{0} a_{j-k} a_{-k}$$

$$= \sum_{n=0}^{\infty} a_{j+n} a_n$$

$$= \sum_{n=0}^{\infty} \beta^{2n+j}$$

$$\Rightarrow \quad a_i = \beta^i.$$

Hence, the transform becomes

$$a(s) = 1/(1-\beta s)$$

$$a_i^{-1} = \begin{cases} 1 & i=0 \\ -\beta & i=1 \\ 0 & i>1. \end{cases}$$

This is the analog of the continuous O–U process as shown in Figure 7.1.

Now, from the Wiener point of view,

$$c(\lambda) = \sum_{\mu=0}^{\infty} k_\mu \, o(\lambda - \mu), \quad \lambda \geq 0$$

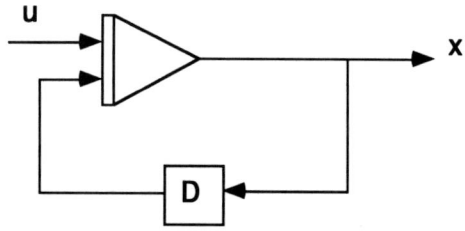

Figure 7.1: Block diagram of the O–U process.

1. Classical Steady-State Filtering

is a convolution:

$$\hat{c} = P\left(\hat{k} * a * \bar{a}\right).$$

Define an operator and propose the solution as follows:

$$\hat{k} = \left[P\left(c * \bar{a}^{-1}\right)\right] * a^{-1},$$

where

$$\hat{k} = \begin{cases} 0 & i < 0 \\ k_i & i \geq 0. \end{cases}$$

This formula is the time domain analog of the classical Wiener spectral factorization formula,

$$F(i\omega) = \frac{1}{(S+N)}\left(\frac{S}{(S+N)^-}\right)_\oplus,$$

where S and N are, respectively, the signal and noise spectral densities; see [19] and [42].

Note that the operator P takes a sequence b and gives $\hat{b}$; it maps

$$l^1(-\infty, \infty) \to h^1(0, \infty)$$

(i.e., h^1 is that subset of l^1 whose sequences have negative indices zero). The convolution operation, $(*)$, maps

$$h^1(0, \infty) * h^1(0, \infty) \to h^1(0, \infty).$$

Now, look at $\hat{k} * o$ for $\hat{c} = P(\hat{k} * o)$. This implies the following:

$$\begin{aligned} \hat{k} * o &= \left[P\left(c * \bar{a}^{-1}\right)\right] * \bar{a} \\ &= c - \left[P^\perp\left(c * \bar{a}^{-1}\right)\right] * \bar{a} \end{aligned}$$

but the $\bar{a}^{-1}$ must vanish for all negative indices; hence

$$P\left(\hat{k} * o\right) = P(c)$$
$$= \hat{c}.$$

This shows analytically that the solution resides in a Banach subalgebra, h^1.

Example 7.2

$$b = (\cdots 0, 1, \lambda\lambda^2, \cdots) \in h^1, \quad |\lambda| < 1$$

$$o = r\delta + b * \bar{b} \quad \text{(i.e., noise is white)}.$$

Aside: note that

$$\sum_{j=-\infty}^{\infty} \frac{\lambda^j s^j}{1-\lambda^2} = \frac{1}{1-\lambda^2} \left\{ \underbrace{\frac{1}{1-\lambda s}}_{0 \to \infty} + \underbrace{\frac{\lambda/s}{1-\lambda/s}}_{-1 \to -\infty} \right\}.$$

Hence, we can factor as follows:

$$o(s) = r + \frac{1}{1-\lambda^2}\left\{\frac{1-\lambda/s+\lambda/s-\lambda^2}{(1-\lambda s)(1-\lambda/s)}\right\}$$

$$o(s) = r + \frac{1}{(1-\lambda s)(1-\lambda/s)},$$

i.e., represent the output sequence as the convolution of two time sequences

$$o(s) = r\left(\frac{(1-\lambda s)(1-\lambda/s)+1/r}{(1-\lambda s)(1-\lambda/s)}\right),$$

which leads to the spectral factor

$$\sqrt{r}\,\frac{\alpha - \beta s}{1 - \lambda s}$$

1. Classical Steady-State Filtering

and

$$\alpha^2 + \beta^2 = \lambda^2 + 1 + \frac{1}{r}$$

$$\beta = \lambda,$$

where the negative sign in the numerator assures positive factors for s^j.

Hence,

$$a^{-1} \sim \frac{1}{\alpha\sqrt{r}} \frac{1 - \lambda s}{1 - \frac{\beta}{\alpha}s}$$

and with $c = \frac{1}{(1-\lambda s)(1-\lambda/s)}$ the convolution yields

$$c * \bar{a}^{-1} = \frac{1}{(1-\lambda s)(1-\lambda/s)} \frac{\frac{1}{\alpha\sqrt{r}}(\frac{1-\lambda}{s})}{(1 - \frac{\beta}{\alpha s})}$$

$$= \frac{1}{\sqrt{r}(\alpha - \beta/s)(1 - \lambda s)}$$

$$= \frac{1}{\sqrt{r}} \frac{1}{\alpha - \beta/s - \alpha\lambda s + \beta\lambda}$$

$$= \frac{1}{\sqrt{r}} \frac{A}{(1 - \lambda s)} + \frac{B}{(\alpha - \beta/s)}.$$

Hence

$$A = \frac{1}{\alpha - \beta\lambda}$$

$$\Rightarrow P(c * \bar{a}^{-1})_j = \begin{cases} 0 & j < 0 \\ \lambda^j / [r^{1/2}(\alpha - \beta\lambda)] & j \geq 0 \end{cases}$$

or in transfer function form

$$\frac{1}{\sqrt{r}} \frac{1}{(1 - \lambda s)(\alpha - \beta\lambda)} \cdot \frac{1}{\alpha\sqrt{r}} \frac{(1 - \lambda s)}{(1 - \beta s/\alpha)}, \qquad \alpha\beta = \lambda,$$

so that

$$\hat{k} = \frac{1}{\alpha r} \frac{1}{(\alpha - \beta\lambda)(1 - \beta s/\alpha)}$$

$$= \frac{1}{r(\alpha^2 - \lambda^2)(1 - \beta s/\alpha)}$$

$$\Rightarrow \quad k_j = \begin{cases} 0 & j < 0 \\ \frac{1}{r(\alpha^2 - \lambda^2)} \left(\frac{\beta}{\alpha}\right)^j & j \geq 0. \end{cases}$$

Note that this is based on the fact that rational spectra result from passing white noise through a linear finite-dimensional filter.

This same example can be more easily resolved using sequential filtering theory as follows. First, find the equilibrium solution of the discrete Riccati equation

$$p_{n+1} = \lambda^2 \frac{p_n r}{p_n + r} + q,$$

which is

$$\bar{p} = -f + \sqrt{f^2 + rq},$$

where

$$f = \frac{(1 - \lambda^2)r - q}{2}.$$

The optimal gain is then

$$\frac{\lambda \bar{p}}{\bar{p} + r}$$

and

$$\hat{x}_{n+1} = \lambda \hat{x}_n + \frac{\lambda \bar{p}}{\bar{p} + r}(z_n - \hat{x}_n),$$

1. Classical Steady-State Filtering

where

$$\hat{x}_n = E(x_n | z_{n-1}, \ldots, z_0).$$

The next lecture will show the multi-input analog of this lecture but with the more powerful Riccati equation method, and we will introduce the Bass theorem.

Lecture 8

A Priori Bounds

1 A Priori Bounds for the Riccati Equation

We want to study the equation

$$P_{n+1} = \phi_{n+1}\left(P_n - P_n H'_n (H_n P_n H'_n + R_n)^{-1} H_n P_n\right)\phi'_{n+1} + G_{n+1} Q_n G'_{n+1},$$

where $P_{n_0} = \Gamma$. Recall that P_{n+1} represents the prediction error covariance at time $n+1$ given data to time n, while $P_n - P_n H'_n (H_n P_n H'_n + R_n)^{-1} H_n P_n$ represents the filter error covariance, the error covariance matrix of the signal process x at time n given data to time n. We have and will assume the matrix Φ_n is invertible for all n. When the processes arise from sampling of continuous time diffusion processes this assumption is fulfilled. Define the mapping τ_{n+1} as

$$\tau_{n+1}(P_n) \triangleq \phi_{n+1}\left(P_n - P_n H'_n (H_n P_n H'_n \right.$$
$$\left. + R_n)^{-1} H_n P_n\right)\phi'_{n+1} + G_{n+1} Q_n G'_{n+1}.$$

Define the set $M_d(R)$ as the set of all $d \times d$ real-entried matrices: $M_d(R)$ is a subset of R^{d^2}. Next, define a subset of $M_d(R)$ consisting of all symmetric $d \times d$ real-entried matrices:

$$SM_d(R) = \{A \in M_d(R) \mid A = A'\}.$$

On this set, define a subset C containing all positive semidefinite $d \times d$ real-entried matrices:

$$C = \{A \in SM_d(R) \mid A \text{ is p.s.d.}\}.$$

The set C is a cone.

Definition 8.1 C *is a cone iff* $C + C \subseteq C$ *and* $\lambda C \subseteq C$ $\forall \lambda > 0$.

For every cone there is a partial ordering: if $A \in C$ and $B \in C$, then $A \geq B$ iff $A - B \in C$.

NB: The existence of a partial ordering on a set does not imply that all elements are comparable.

Remark: Recall that the 2×2 matrix $\begin{pmatrix} a & b \\ b & c \end{pmatrix}$ is p.s.d. iff $a \geq 0$, $c \geq 0$, and $ac - b^2 \geq 0$.

Next, recall the definition of the Moore-Penrose pseudo inverse. It says that if $A \in M_d(R)$, then $A^{\#}$ is the unique matrix that satisfies

1. $AA^{\#}A = A$

2. $A^{\#}AA^{\#} = A^{\#}$

3. $(AA^{\#})' = AA^{\#}$

4. $(A^{\#}A)' = A^{\#}A$.

The matrix $P = AA^{\#}$ is a projection onto the range of A and is *idempotent* (i.e., $P^2 = P$). For the geometric interpretation of the Moore-Penrose pseudo inverse, see [41].

Definition 8.2 (Duffin) *If* $A, B \in SM_d(R)$, *then*

$$(A : B) \triangleq A(A+B)^{\#}B.$$

1. A Priori Bounds for the Riccati Equation

Remark: $(A:B)$ is called the parallel resistance, from the circuit property for scalar A and B.

Lemma 8.1 *If $A_i \in M_{\ell,k}(R)$, $i = 1, \ldots, n$, then*

$$A_i \left(\sum_{j=1}^{n} A'_j A_j \right) \left(\sum_{j=1}^{n} A'_j A_j \right)^{\#} = A_i$$

for $1 \leq i \leq n$.

The proof is left to the reader as an exercise.

Remark 1: Multiplying both sides of the above equation by A'_i, we obtain

$$A'_i A_i \left(\sum_{j=1}^{n} A'_j A_j \right) \left(\sum_{j=1}^{n} A'_j A_j \right)^{\#} = A'_i A_i \quad \text{for } 1 \leq i \leq n.$$

Remark 2: By property 3 of the Moore-Penrose pseudo inverse,

$$A_i \left(\sum_{j=1}^{n} A'_j A_j \right)^{\#} \left(\sum_{j=1}^{n} A'_j A_j \right) = A_i \quad \text{for } 1 \leq i \leq n.$$

Lemma 8.2 *For $A, B \in C$,*

$$A \geq B \implies \dot{\tau}_n(A) \geq \dot{\tau}_n(B),$$

where $\dot{\tau}$ is the same as τ except for a pseudo inverse in place of the inverse.

Remark: This lemma says that the mapping $\dot{\tau}$ preserves order.

Proof: Consider $\|x\|^2_{\dot{\tau}_n(A)}$. It will satisfy the quadratic form

$$\|x\|^2_{\dot{\tau}_n(A)} = \min_{r \in R^s} \left\{ \|\phi'_n x + H'_{n-1} r\|^2_A + \|r\|^2_{R_{n-1}} + \|x\|^2_{C_n} \right\},$$

where $C_n = G_n Q_{n-1} G'_n$. This is quadratic in r. The first variation in r is

$$\delta r' \big(H_{n-1} A \phi'_n x + (H_{n-1} A H'_{n-1} + R_{n-1}) r \big)$$

and the second variation in r is

$$\delta r'(H'_{n-1}AH_{n-1} + R_{n-1})\delta r.$$

Since the second variation is ≥ 0, it is p.s.d. Let r_0 be a minimizing value of r, which exists as the function is quadratic. It must satisfy

$$\delta r'\left(H_{n-1}A\phi'_n x + (H_{n-1}AH'_{n-1} + R_{n-1})r_0\right) = 0.$$

Since δr is arbitrary,

$$(H_{n-1}AH'_{n-1} + R_{n-1})r_0 = -H_{n-1}A\phi'_n x.$$

We want to show that there is a solution to this equation. Let

$$r_* \triangleq -K^\# H_{n-1} A\phi'_n x,$$

where

$$K \triangleq H_{n-1}AH'_{n-1} + R_{n-1}.$$

Then, since A is p.s.d., we can take its square root and write

$$Kr_* = -KK^\# H_{n-1} A^{1/2}(A^{1/2}\phi'_n x).$$

But, by Lemma 1.1, this reduces to

$$Kr_* = -H_{n-1}A^{1/2}(A^{1/2}\phi'_n x).$$

Thus, $r_0 = r_*$. Substituting for r and some algebra gives the claimed result.

We have shown that C is an invariant set for $\dot\tau_n$: $\dot\tau_n(C) \subseteq C$.

Denote τ_n as $\Pi(n, \Gamma, n_0)$. Then we have shown

$$\Pi(n_0 + i, 0, n_0) \geq \Pi(n_0 + i, 0, n_0 + 1) \geq 0$$

1. A Priori Bounds for the Riccati Equation

since

$$\Pi(n_0 + i + 1, 0, n_0) = \Pi\Big(n_0 + i + 1, \Pi(n_0 + 1, 0, n_0), n_0 + 1\Big)$$

$$\geq \Pi(n_0 + i + 1, 0, n_0 + 1).$$

Thus $\Pi(t, 0, \sigma)$ increases as σ decreases.

Theorem 8.1 *If $0 \leq \alpha_n \leq \beta$ and $\alpha_n \uparrow$, then $\exists\, \alpha \ni \alpha_n \to \alpha$ (a bounded monotone sequence converges).*

Theorem 8.2 (Riesz) *If $A_n \in C$ and $\exists\, \beta \ni A_n \leq \beta I$ and $A_n \leq A_{n+1}\ \forall n$ (i.e., A_n is bounded, monotone), then $\exists\, A \ni A_n \to A$ as $n \to \infty$.*

Proof: We know that $x' A_n x$ converges to α by Theorem 1.1. Define

$$a_{ii} \triangleq \lim_{n \to \infty} e_i' A_n e_i,$$

where the e_i are the elementary unit vectors. Then

$$(e_i + e_j)' A_n (e_i + e_j) = e_i' A_n e_i + 2 e_j' A_n e_i + e_j' A_n e_j \to \beta_{ij}.$$

All terms converge. Thus $e_j' A_n e_i$ converges to a_{ij}. Thus, for arbitrary x and y,

$$\left(\sum_{i=1}^n x_i e_i\right)' A_n \left(\sum_{j=1}^n y_j e_j\right) = \sum_{i,j=1}^n x_i y_j e_i' A_n e_j \to \sum_{i,j=1}^n x_i y_j a_{ij}.$$

Q.E.D.

Recall that

$$\|x\|_{\tau_n(A)}^2 = \min_{r \in R^s} \left\{ \|\phi_n' x + H_{n-1}' r\|_A^2 + \|r\|_{R_{n-1}}^2 + \|x\|_{C_n}^2 \right\}.$$

Let's introduce something from control theory. Think of x as the state and choose a feedback controller (deadbeat control) that successively takes out part of x. We are trying to find an upper bound on $\Pi(n, \Gamma, n_0)$ as $\Gamma \to \infty$. We want $n - n_0$ to be big. The model is

$$x_{n+1} = \phi_{n+1} x_n + G_{n+1} u_n,$$

where the initial state is a constant, $x_{n_0} = c$. We can drive the state to 0 in one step for x in the range of $(\phi_{n_0+1}^{-1} G_{n_0+1})$; i.e., for

$$x_{n+1} = \phi_{n_0+1}^{-1} G_{n_0+1} d.$$

We can drive the state to 0 in two steps for x in the range of

$$\left((\phi_{n_0+2} \phi_{n_0+1})^{-1} G_{n_0+2}, \; \phi_{n_0+1}^{-1} G_{n_0+1} \right)$$

as follows:

$$x_{n_0} = \phi_{n_0+1}^{-1} G_{n_0+1} d + \phi_{n_0+1}^{-1} \phi_{n_0+2}^{-1} G_{n_0+2} e$$

$$x_{n_0+1} = \phi_{n_0+2}^{-1} G_{n_0+2} e.$$

Define

$$A(n+d) = \phi(n_0+d, n_0+d+1), \ldots, \phi(n_0+1, n_0).$$

We can drive the state to 0 in d steps for x in the range of

$$\left(A^{-1}(n_0+1) G_{n_0+1}, \ldots, A^{-1}(n_0+d) G_{n_0+d} \right).$$

Assume this system is uniformly completely controllable (u.c.c.). This is a deadbeat controller, a feedback controller that computes the expansion of the state in terms of the columns of these composite matrices. This

1. A Priori Bounds for the Riccati Equation

assumption allows us to force anything to the origin in d steps. Recall

$$\text{u.c.c.} \Rightarrow \text{Rank}\left[A^{-1}(n_0+1)G_{n_0+1}, \ldots, A^{-1}(n_0+d)G_{n_0+d}\right] = d.$$

Now, let's do the dual thing for the filter. We will need uniform complete observability (u.c.o.). Consider

$$\|x\|^2_{\Pi(n,\Gamma,n_0)} = \min_{r \in R^s}\left\{\|\phi'_n x + H'_{n-1}r\|^2_{\tau_n(\Pi(n-1,\Gamma,n_0))} + \|r\|^2_{R_{n-1}} + \|x\|^2_{C_n}\right\}.$$

Any choice of r gives us an upper bound. We choose the deadbeat sequence. Note that the deadbeat sequence of rs yields a bound independent of Γ. Expand the initial state as follows:

$$x_0 = \sum_{i=1}^{d} A^{-1}(n_0+i)H'_{n_0+i}a_i.$$

Then the deadbeat sequence is

$$r_k = -a_k.$$

Assume uniform complete observability, $R_k \leq \alpha I$, $Q_k \leq \alpha I$, and $\|A(i)\| \leq K$, a constant matrix. Then

$$\|x\|^2_{\Pi(m+d,\Gamma,m)} \leq M\|x\|^2 I \quad \forall\, \Gamma, m \tag{8.1}$$

for some constant matrix M. Now use the semigroup property to get the bound for any $n > n_0 + d$:

$$\Pi(n,\Gamma,n_0) = \Pi\left(n, \Pi(n-d+1,\Gamma,n_0), n-d+1\right) \leq M.$$

This holds as long as the indices are far enough apart. These bounds are called *a priori bounds*.

Note that $\Pi(n,0,n_0)$ increases as n_0 decreases. We have shown that we can find a bound for the first d matrices. Combining this with Equation

(8.1), we see that the whole sequence is bounded. Note that Equation (8.1) is independent of Γ and involved only d steps.

Lemma 8.3

$$\lim_{n_0 \to -\infty} \Pi(n, \Gamma, n_0) = \bar{P}(n) \stackrel{\Delta}{=} \lim_{n_0 \to -\infty} \Pi(n, 0, n_0).$$

Theorem 8.3

$$\Gamma > 0 \quad \Rightarrow \quad \Pi(n, \Gamma, n_0) > 0.$$

This follows by induction.

Assume the system is u.c.c. and u.c.o., and assume P_{n+1} can be expressed in the form

$$P_{n+1} = Y_{n+1} X_{n+1}^{-1}.$$

This is valid provided X_n is invertible. Using the Schur relations,

$$P_{n+1} = \phi_{n+1}(P_n^{-1} + H_n' R_n^{-1} H_n)^{-1} \phi_{n+1}' + G_{n+1} Q_n G_{n+1}'.$$

Assume $0 < \alpha_1 I < R_n < \alpha_2 I$ and let $M_n = H_n' R_n^{-1} H_n$ and $C_{n+1} = G_{n+1} Q_n G_{n+1}'$. Then

$$P_{n+1} = \phi_{n+1}(X_n Y_n^{-1} + M_n)^{-1} \phi_{n+1}' + C_{n+1}$$

$$Y_{n+1} X_{n+1}^{-1} = \phi_{n+1} Y_n (X_n + M_n Y_n)^{-1} \phi_{n+1}' + C_{n+1}$$

$$= \left[\phi_{n+1} Y_n + C_{n+1} \phi_{n+1}'^{-1}(X_n + M_n Y_n) \right] \phi_{n+1}'^{-1}(X_n + M_n Y_n).$$

This can be put in the following form:

$$\begin{pmatrix} X_{n+1} \\ Y_{n+1} \end{pmatrix} = \begin{pmatrix} \phi_{n+1}'^{-1} & \phi_{n+1}'^{-1} M_n \\ C_{n+1} \phi_{n+1}'^{-1} & \phi_{n+1} + C_{n+1} \phi_{n+1}'^{-1} M_n \end{pmatrix} \begin{pmatrix} X_n \\ Y_n \end{pmatrix}$$

2. Information and Filtering

$$= S\begin{pmatrix} X_n \\ Y_n \end{pmatrix}.$$

Thus we can produce a solution of the Riccati equation so long as $X_n > 0$ in analogy with the scalar linearization presented in Lecture 4. When $\det(X_n) = 0$, we have a conjugate point, it will be demonstrated later that under the conditions of u.c.c and u.c.o. X_n is nonsingular. Note that S is symplectic. Also note that P^{-1} satisfies an equation similar to the preceding equation. Thus

$$mI \geq \Pi^{-1}(n_0 + d, \Gamma, n_0).$$

Assuming the system is u.c.c., we get

$$\Pi(n_0 + d, \Gamma, n_0) \geq mI.$$

Thus we are positive definite after d steps; we have been attracted inside the cone.

2 Information and Filtering

Assume that η and ζ are zero-mean Gaussian random vectors with positive definite joint covariance. It follows from Theorem 1.3 of Lecture 1 that

$$E\eta|\zeta = E\eta\zeta'(E\zeta\zeta')^{-1}\zeta, \tag{8.2}$$

$$E(\eta - E\eta|\zeta)(\eta - E\eta|\zeta)' = E\eta\eta' - E\eta\zeta'(E\zeta\zeta')^{-1}E\zeta\eta'. \tag{8.3}$$

The Shannon joint information $I(\eta, \zeta)$ defined by

$$I(\eta, \zeta) = E \log\left(\frac{p^{\eta,\zeta}(x,y)}{p^\eta(x)p^\zeta(y)}\right),$$

where $p^{\eta,\zeta}(x,y)$ is the joint density of η and ζ and $p^\eta(x)$ and $p^\zeta(y)$ are the marginal densities. The joint information is easily shown to be

$$I(\eta,\zeta) = \log\left(\frac{\det\begin{pmatrix} A & B \\ B' & C \end{pmatrix}}{\det(A)\det(C)}\right),$$

where $A = E\eta\eta'$, $B = E\eta\zeta'$, and $C = E\zeta\zeta'$, where $\det(R)$ is the determinant of the matrix R and will be denoted by $|R|$ subsequently. The Shur relations of Lecture 1 imply that

$$\begin{vmatrix} A & B \\ B' & C \end{vmatrix} = |C||A - BC^{-1}B'|$$

so that

$$I(\eta,\zeta) = \log\left(\frac{|A - BC^{-1}B'|}{|A|}\right). \tag{8.4}$$

Recall the variables of the linear Gaussian filtering problem x_n and z^n. In terms of these variables and their variants we have the following theorem.

Theorem 8.4 *For the linear filtering problem:*

$$I(x_n; z^{n-1}) = \log\left(\frac{|P_n|}{|S_n|}\right) = I(x_n; Ex_n|z^{n-1}), \tag{8.5}$$

where P_n is the one-step prediction error covariance matrix.
Further,

$$I(x_n; z^n) = \log\left(\frac{|\Sigma_n|}{|S_n|}\right) = I(x_n; Ex_n|z^n) \tag{8.6}$$

with $\Sigma_n = P_n - P_n H_n'(H_n P_n H_n + R_n)^{-1} H_n P_n$, the filtering error covariance matrix.

Proof. In view of (8.2), (8.3), (8.4), and the relation

$$E(E\eta|\zeta)\zeta) = E\eta\zeta,$$

the assertions follow.

Remark: Notice that the first assertion of the theorem says that $E\eta|\zeta$ carries all the information contained in z^{n-1} about x_n.

Corollary 8.1 *For the linear filtering problem*

$$I(x^n; z^{n-1}) = \log\left(\frac{|\mathcal{P}_n|}{|\mathcal{S}_n|}\right) \tag{8.7}$$

and

$$I(x^n; z^n) = \log\left(\frac{|\mathcal{T}_n|}{|\mathcal{S}_n|}\right), \tag{8.8}$$

where $\mathcal{S}_n = Ex^n x^n$, $\mathcal{P}_n$ *is the covariance matrix of* $x^n - Ex^n|z^{n-1}$, *and* $\mathcal{T}_n$ *is the covariance matrix of* $x^n - Ex^n|z^n$.

Notice that the diagonal elements of $\mathcal{S}_n$ and $\mathcal{T}_n$ are the interpolation errors. Gelfand and Yaglom [34] proved the continuous analog of the theorem proved here, while Pinsker gave the proof of the corollary [35]. An interesting design method for reduced state dimension filters based on information theory is given in [38]. Results for the nonlinear filtering problem are detailed in [39]. Sources for related information are [37], [40], and [36].

3 Nonlinear Systems

Suppose the system satisfies a nonlinear difference equation:

$$\mathbf{x}_{n+1} = f(\mathbf{x}_n, n).$$

Let $\bar{\mathbf{x}}(n)$ denote the equilibrium solution. Without loss of generality, we assume $\bar{\mathbf{x}}(n) = 0$.

Definition 8.3 *This system is uniformly asymptotically stable (u.a.s.) iff* $\forall \epsilon > 0 \; \exists \delta \ni$

$$\|\mathbf{x}(n_0) - \bar{\mathbf{x}}(n_0)\| < \delta \quad \Rightarrow \quad \|\mathbf{x}(n) - \bar{\mathbf{x}}(n)\| < \epsilon$$

$\forall\ n > n_0$ and $\mathbf{x}(n) \to \bar{\mathbf{x}}(n)$ as $n \to \infty$.

Theorem 8.5 (Lyapunov) *Suppose there exists $V(\mathbf{x}, n)$ scalar, and $\bar{\mathbf{x}}(n)$ such that*

- $V\big(\mathbf{x}(n), n\big) > 0, \quad \mathbf{x} \neq \bar{\mathbf{x}}, \quad V\big(\bar{\mathbf{x}}(n), n\big) = 0.$

- *Assume there exist α_1 and α_2 such that*

$$\alpha_1 \|\mathbf{x} - \bar{\mathbf{x}}\|^2 \leq V(\mathbf{x}, n) \leq \alpha_2 \|\mathbf{x} - \bar{\mathbf{x}}\|^2.$$

- $V\big(\bar{\mathbf{x}}(n+1), n+1\big) - V\big(\bar{\mathbf{x}}(n), n\big) \leq -\gamma \|\mathbf{x} - \bar{\mathbf{x}}\|^2$

for $\alpha_1\ \alpha_2$ and γ then $\bar{\mathbf{x}}$ is uniformly asymptotically stable.

Think of V as the energy function; it vanishes along the equilibrium solution.

Proof: See [31], [32], or [33].

Using results from this lecture, we will prove the existence of a unique positive definite equilibrium point for the Riccati equation. We will also prove that the closed loop filter is u.a.s.

Lecture 9

Asymptotic Theory

1 Applications of the Theory of Filtering

Let's look at the map

$$\Delta_n(A) = \phi_n A \phi_n' - K_n \left(H_{n-1} A H_{n-1}' + R_{n-1} \right)^{\#} K_n' + G_n G_n'$$

$$K_n = \phi_n A H_{n-1}' + G_n L_{n-1}'$$

where

$$R_{n-1} \geq L_{n-1}' L_{n-1}.$$

This form can arise when we look at a block of observations.

Example:

$$x_{n+1} = \phi_{n+1} x_n + G_{n+1} u_n$$

$$y_n = H_n x_n + \nu_n.$$

Make a new observation process:

$$\begin{pmatrix} y_n \\ y_{n+1} \end{pmatrix} = \begin{pmatrix} H_n x_n \\ H_{n+1} x_{n+1} \end{pmatrix} + \begin{pmatrix} \nu_n \\ \nu_{n+1} \end{pmatrix}$$

$$= \begin{pmatrix} H_n x_n \\ H_{n+1}(\phi_{n+1} x_n + G_{n+1} u_n) \end{pmatrix} + \begin{pmatrix} \nu_n \\ \nu_{n+1} \end{pmatrix}$$

$$= \begin{pmatrix} H_n \\ H_{n+1}\phi_{n+1} \end{pmatrix} x_n + \begin{pmatrix} \nu_n \\ \nu_{n+1} + H_{n+1}G_{n+1}u_n \end{pmatrix}$$

$$x_{n+2} = (\phi_{n+2}\phi_{n+1})x_n + \phi_{n+2}G_{n+1}u_n + G_{n+2}u_{n+1}.$$

The cross-correlation between observation noise and plant noise gives rise to a term like $G_n L'_{n-1}$ in K_n:

$$L'_{n-1} = E\{u_{n-1}\nu'_{n-1}\}.$$

We can eliminate the cross-correlation in the general case as follows.

Theorem 9.1 *Suppose* $R^{\#}_{n-1} R_{n-1} L_{n-1} = L_{n-1}$. *Then for* $A \in \mathcal{C}$,

$$\Delta_n(A) = T_n(A),$$

where

$$T_n(A) = \psi_n S_n(A)\psi'_n + G_n(I - L'_{n-1}R^{\#}_{n-1}L_{n-1})G'_{n-1}$$

and

$$S_n(A) = A - AH'_{n-1}(H_{n-1}AH'_{n-1} + R_{n-1})^{\#} H_{n-1}A.$$

The dynamics matrix has changed:

$$\psi_n = \phi_n - GL'_{n-1}R^{\#}_{n-1}H_{n-1}.$$

The factor in parentheses in the preceding equation for $T_n(A)$ is positive semi-definite when the equation arises from block processing of data, which follows from the Schwarz inequality.

Proof

$$\Delta_n(A) = \inf_u \frac{1}{2} \left(||u||_{R_n}^2 + ||G_n c||^2 + 2(u, L_{n-1} G'_n c) + ||\phi'_n c + H'_{n-1} u||_A^2 \right)$$

$$= \inf_u \frac{1}{2} \left(||u + R_{n-1}^{\#} L_{n-1} G'_n c||_{R_n}^2 + ||G_n c||_{I - L'_{n-1} R_{n-1}^{\#} L_{n-1}}^2 \right.$$

$$\left. + ||\phi'_n c + H'_{n-1} u||_A^2 \right)$$

with

$$v = u + R_{n-1}^{\#} L_{n-1} G'_n c$$

$$\Delta_n(A) = \inf_v \frac{1}{2} \left(||v||_{R_n}^2 + ||G_n c||_{I - L'_{n-1} R_{n-1}^{\#} L_{n-1}}^2 + ||\psi'_n c + H'_{n-1} u||_A^2 \right)$$

where

$$\psi_n = \phi_n - G_n L_{n-1} R_{n-1}^{\#} H_{n-1}.$$

The last equation implies the result.

Consider the general case of block processing. We assume stationarity, so there will be no subscripts. From Eq. (2.7),

$$T(A) = \phi'_* \{ A - AH'_*(H_* AH'_* + R_*)^{\#} H_* A \} \phi_* + G_* G'_*.$$

Apply T k times:

$$T^k(A) = \phi^k \{ A - (AH' + \phi^{-k} GL')(HAH' + R + LL')^{\#}(HA + LG'\phi'^{-k}) \} \phi'^k + GG'.$$

Define the "system moments":

$$u_i = H_* \phi'_* G_*.$$

Then

$$L = \begin{bmatrix} 0 & \cdots & & & \\ u_0 & 0 & \cdots & & \\ u_1 & u_0 & 0 & \cdots & \\ \vdots & \vdots & \vdots & \ddots & \\ u_{k-2} & u_{k-3} & \cdots & u_0 & 0 \end{bmatrix}$$

$$R = \begin{bmatrix} R_* & & \\ & \ddots & \\ & & R_* \end{bmatrix} \qquad H = \begin{bmatrix} H_* \\ H_*\phi_* \\ \vdots \\ h_*\phi_*^{k-1} \end{bmatrix}$$

$$G = \begin{bmatrix} \phi_*^{k-1} G_*, \cdots, G_* \end{bmatrix}.$$

Define M by

$$H\phi_*^{-k} G = M + L.$$

2 Asymptotic Theory of the Riccati Equation

Let m be chosen equal to the maximum of the intervals of uniform complete controllability (u.c.c.) and that of uniform complete observability (u.c.o), we assume throughout this section that both conditions hold. Consider the block-processed Riccati equation

$$W(k, \Gamma, l_0, p) = \Pi(km + p, \Gamma, l_0 m + p) \quad p = 0, 1..., m-1.$$

From the last theorem, W satisfies a Riccati equation of the form

$$W(k+1) = A_{k+1}(W(k) - W(k)C_k'(C_k W(k) C_k' + N_k)^\# C_k W(k)) A_{k+1}'$$
$$+ T_{k+1} T_{k+1}',$$

2. Asymptotic Theory of the Riccati Equation

where $A_\bullet$, $C_\bullet$, $T_\bullet$, and $N_\bullet$, depend on p and are expressible in terms of the original system matrices by the transformations of block processing and removing input-output cross-correlation by the previous theorem. In the following we suppress p in the notation, but the system matrices are p-dependent.

Theorem 9.2 (A Priori Bounds) *For all $\Gamma \in \mathcal{C}$ and $k \geq 1$*

$$T_{n_0+1}T'_{n_0+1} \leq W(k+1, \Gamma, l_0, p) \leq A_{k+1}(C'_k N_k^{-1} C_k)^{-1} A'_{k+1}$$

and there exist positive constants l and u such that

$$lI \leq W(k+1, \Gamma, l, 0, p) \leq uI.$$

<u>Proof</u>
The second set of inequalities follows from the first, in view of u.c.c. and u.c.o. The first inequality is immediate from the equation for W and the fact that the Riccati iteration preserves order in $\mathcal{C}$. Hence $W(k+1)$ is invertible for $k \geq 1$. However, the Schur relations show that $M(k) = W(k)^{-1}$ satisfies

$$M(k+1) = ((A_{k+1}(M(k) + C'_k N_k^{-1} C_k)^{-1} A'_{k+1} + T_{k+1} T'_{k+1})^{-1}$$

or

$$M(k+1) \geq A_{k+1}^{-1}{'} C'_k N_k^{-1} C_k A_{k+1}^{-1}$$

and this implies the first upper bound for $W(k+1)$.
The closed-loop filter

$$\bar{A}(W)_{k+1} = A_{k+1} - W(k)C'_k(C_k W(k) C_k + N_k)^{-1} C_k$$

is u.a.s. This follows since $\|x\|^2_{W(k)^{-1}}$ is a Lyapounov function for the linear system of the unforced closed-loop system. Notice that the a priori bounds are essential to give the Lyapounov property. Notice that

$$\lim W(k, 0, l_0, p) \text{ as } l_0 \to -\infty$$

exists and equals

$$\bar{P}(km + p) = \lim \Pi(km + p, 0, n_0) \text{ as } n_0 \to -\infty$$

as both are bounded monotone sequences that converge by the Riesz theorem, and the limits coincide as the W sequence is a subsequence of the Π sequence. The following theorem shows that $\bar{P}$ is the unique attractor for W in the cone $\mathcal{C}$.

Theorem 9.3 *For arbitrary $\Gamma \in \mathcal{C}$*

$$\bar{P}(km + p) = \lim W(k, \Gamma, l_0, p) \text{ as } l_0 \to -\infty.$$

Proof

For Γ_1 and Γ_2 any elements of $\mathcal{C}$ and any p: $0 \leq p \leq m - 1$, let

$$W_1(k) = W(k, \Gamma_1, l_0, p)$$

and

$$W_2(k) = W(k, \Gamma_2, l_0, p).$$

The difference $W_1(k) - W_2(k) = \Delta(k)$ satisfies

$$\Delta(k+1) = \bar{A}(W_1)_{k+1} \Delta(k) \bar{A}(W_2)'_{k+1},$$

which is bilinear, and the two system matrices $\bar{A}(\bullet)$ are the closed-loop dynamics of u.a.s. linear systems since $||x||^2_{W_i^{-1}}$ i=1,2 are Lyapunov functions

2. Asymptotic Theory of the Riccati Equation

for the linear systems. The equation for Δ then implies

$$||\Delta(k)|| \leq ce^{-v(k-l_0)}$$

or since $W(k, 0, l_0, p)$ converges to $\bar{P}(km + p)$ all solutions starting from $\mathcal{C}$ converge to the same limit.

Now the sequence $\Pi(n, \infty, n_0)$ is well defined as Π^{-1} satisfies a Riccati equation. Denote the solution of this latter equation as $\Sigma(n, \Gamma, n_0)$ and define $\Pi(n, \infty, n_0)$ as $[\Sigma(n, 0, n_0)]^{-1}$. As $n_0 \to -\infty$ the sequence $\Pi(n, \infty, n_0)$ is monotone and hence converges to $''P(n)$. But $''P(n) = \bar{P}(n)$ since for each n, $n = km + p$ for unique k and p, so that $W(k, \infty, l_0, p)$ converges as $l_0 \to -\infty$ to $\bar{P}(n)$ by the previous theorem, while on the other hand this sequence is a subsequence of $\Pi(n, \infty, n_0)$ for $n_0 \to -\infty$ and hence must converge to $''P(n)$ so that $''P(n) = \bar{P}(n)$.

Theorem 9.4 *For all* $\Gamma \in \mathcal{C}$

$$\bar{P}(n) = \lim_{n_0 \to -\infty} \Pi(n, \Gamma, n_0).$$

Proof
For arbitrary $\Gamma \in \mathcal{C}$, since the Riccati map preserves order

$$\Pi(n, 0, n_0) \leq \Pi(n, \Gamma, n_0) \leq \Pi(n, \infty, n_0).$$

But the extremes of this inequality both converge to $\bar{P}(n)$, which demonstrates the theorem.

Summary:
We have shown that we have asymptotic stability and uniqueness for the Riccati equation. The composite mapping is of considerable interest. To find periodic solutions, we look for the fixed points of the composite map T^n. Each such fixed point generates a periodic solution of period n.

3 Steady-State Solution to Riccati

Let's view things another way. Consider a Riccati equation corresponding to a model where the plant and observation noise are uncorrelated. This can be achieved in general. Recall that we can write P_n in the form

$$P_n = Y_n X_n^{-1}.$$

Complete observability implies that X_n is invertible. Now consider

$$\begin{pmatrix} X_n \\ Y_n \end{pmatrix} = \mathcal{S}_n \begin{pmatrix} X_{n-1} \\ Y_{n-1} \end{pmatrix},$$

where

$$\mathcal{S}_n = \begin{pmatrix} \phi_n'^{-1} & \phi_n'^{-1} M_{n-1} \\ C_n \phi_n'^{-1} & \phi_n + C_n \phi_n'^{-1} M_{n-1} \end{pmatrix}$$

$$M_{n-1} = H_{n-1}' R_{n-1}' H_{n-1}$$

$$C_n = G_n \phi_{n-1} G_n'.$$

$\mathcal{S}_n$ is a $2d \times 2d$ symplectic matrix, which corresponds to the τ_n operator from the previous lecture, which was bilinear, positive definite, and $d \times d$. Thus $\tau_2 \tau_1$ corresponds to $\mathcal{S}_2 \mathcal{S}_1$; i.e., the composition of the bilinear mappings becomes a matrix multiply of symplectic matrices, where the bilinear mappings are linear fractional transformations of the form

$$\frac{\alpha z + \beta}{\gamma z + \delta}.$$

We want to show another way to solve the Riccati equation which is connected to the Weiner theory.

Assume ϕ, G, Q, R, and H are time-independent and the system is u.c.c. and u.c.o., and let

$$\bar{P} = \lim_{t_o \downarrow -\infty} \Pi(t, A, t_o) \qquad A \in \mathcal{C}.$$

3. Steady-State Solution to Riccati

Theorem 9.5 (Bass-Roth) *Assume complete observability and controlability. Then*

$$\det(sI - S) = (-s)^{2d} a(s) a\left(\frac{1}{s}\right),$$

where $a(s)$ has all its roots within the unit circle, and

$$a(s) = \det(sI - \bar{\phi}),$$

where

$$\bar{\phi} = \phi - \phi \bar{P} H'(H \bar{P} H' + R)^{-1} H$$

and

$$[-\bar{P}, I] a(\mathcal{S}) = 0 \;!!$$

Corollary 9.1 (Potter) *Let $e_1, \ldots, e_d$ be the eigenvectors of $\mathcal{S}$ corresponding to the roots of $\det(sI - \mathcal{S})$ outside the unit circle. Also assume complete observability and controlability. Then*

$$\mathcal{S}(e_1 \cdots e_d) = \begin{pmatrix} X \\ Y \end{pmatrix}$$

and

$$\bar{P} = Y X^{-1} \;!!!$$

Proof of Bass-Roth:

$$\begin{aligned} h(s) \;&=\; \det(sI - \mathcal{S}) = \det J' J \det(sI - \mathcal{S}) \\ &=\; \det(sI - J'\mathcal{S}J) = \det(sI - \mathcal{S}'^{-1}) \end{aligned}$$

$$\begin{aligned}
&= \det(sI - \mathcal{S}^{-1}) = \det \mathcal{S}^{-1} \det(sI\mathcal{S} - I) \\
&= \det(I\mathcal{S} - \frac{1}{s}I)s^{2d} \\
&= (-s)^{2d} h(\frac{1}{s}) \quad = (-s)^d \det \mathcal{S}^{-1}.
\end{aligned}$$

Thus, if s is a root, $\frac{1}{s}$ is a root:

$$\begin{pmatrix} I & 0 \\ -\bar{P} & I \end{pmatrix} \begin{pmatrix} \phi'^{-1} & \phi'^{-1}M \\ C\phi'^{-1} & \phi + C\phi'^{-1}M \end{pmatrix} \begin{pmatrix} I & 0 \\ \bar{P} & I \end{pmatrix}$$

$$= \begin{pmatrix} I & 0 \\ -\bar{P} & I \end{pmatrix} \begin{pmatrix} \phi'^{-1} + \phi'^{-1}M\bar{P} & \phi'^{-1}M \\ C\phi'^{-1} + \phi\bar{P} + C\phi'^{-1}M\bar{P} & \phi + C\phi'^{-1}M \end{pmatrix}$$

$$= \begin{pmatrix} \bar{\phi}'^{-1} & \phi'^{-1}M \\ \star & \phi + C\phi'^{-1}M - \bar{P}\phi'^{-1}M \end{pmatrix}.$$

We know

$$P = \phi(P^{-1} + M)^{-1}\phi' + C$$

$$\bar{\phi} = \phi - \phi PH'(HPH' + R)^{-1}H.$$

Thus

$$\bar{\phi}P = \phi(P - PH'(HPH' + R)^{-1}HP) = \phi(P^{-1} + M)^{-1}$$

$$\bar{\phi} = \phi(P^{-1} + M)^{-1}P^{-1} = \phi(I + PM)^{-1}$$

and so

$$\bar{\phi}^{-1} = (I + PM)\phi^{-1}.$$

Hence

$$\star = C\bar{\phi}'^{-1} + \phi\bar{P} - \bar{P}\bar{\phi}'^{-1}$$

3. Steady-State Solution to Riccati

$$= \{C + \phi \bar{P} \bar{\phi}' - \bar{P}\} \bar{\phi}'^{-1}.$$

Then

$$\phi \bar{P} \bar{\phi}' = \phi P(I - MP)^{-1} \phi'$$
$$= \phi (P^{-1} + M)^{-1} \phi';$$

the expression within braces in the previous equation is the Riccati equation, which is identically zero at this stationary solution:

$$C + \bar{\phi} \bar{P} \phi' - \bar{P} = 0.$$

Thus, $\star = 0$. Multiplying this from the left by $\phi'^{-1}M$, we get

$$C\phi'^{-1}M - \bar{P}\phi'^{-1}M = -\bar{\phi} PM = -\phi(I + PM)^{-1}PM$$
$$= \bar{\phi}.$$

We choose $a(s)$ to be the characteristic polynomial of $\bar{\phi}$. Using complete observability and controllability, we can show

$$a(s) = \det(sI - \bar{\phi}),$$

which is the first result of Bass-Roth. To prove the second result, we consider

$$\begin{pmatrix} I & 0 \\ -\bar{P} & I \end{pmatrix} a(s) \begin{pmatrix} I & 0 \\ P & I \end{pmatrix} = \begin{pmatrix} a(\bar{\phi}'^{-1}) & \star \\ 0 & 0 \end{pmatrix},$$

where we have used the fact that $a(\bar{\phi})$ vanishes by Cayley-Hamilton. Let

$$a(s) = \begin{pmatrix} A & B \\ C & D \end{pmatrix}.$$

Then

$$\begin{pmatrix} I & 0 \\ -\bar{P} & I \end{pmatrix} a(s) \begin{pmatrix} I & 0 \\ \bar{P} & I \end{pmatrix} = \begin{pmatrix} A + B\bar{P} & B \\ C - \bar{P}A + (D - \bar{P}B)\bar{P} & D - \bar{P}B \end{pmatrix}$$

so that

$$C - \bar{P}A + (D - \bar{P}B)\bar{P} = 0$$

$$D - \bar{P}B = 0.$$

From this the second Bass-Roth result follows:

$$[-\bar{P}, I] a(\mathcal{S}) = 0.$$

Note that this is the spectral factorization of the filter output.

The dual equation to the preceding one is

$$a(\mathcal{S}^{-1}) \begin{bmatrix} I \\ P \end{bmatrix} = 0.$$

<u>Proof of Potter</u>: We prove the result in the special case of distinct eigenvalues only; the general case is left to the reader.

$$\mathcal{S} e_i = \lambda_i e_i$$

$$\mathcal{S}[e_1 \cdots e_d] = [e_1 \cdots e_d] \begin{bmatrix} \lambda_1 & & \\ & \ddots & \\ & & \lambda_d \end{bmatrix}$$

$$a(\mathcal{S}^{-1})[e_1 \cdots e_d] = [e_1 \cdots e_d] \begin{bmatrix} a\left(\frac{1}{\lambda_1}\right) & & \\ & \ddots & \\ & & a\left(\frac{1}{\lambda_d}\right) \end{bmatrix}.$$

3. Steady-State Solution to Riccati

If λ_i lies outside the unit circle, λ_i^{-1} lies inside. Since

$$a(\mathcal{S}^{-1})\begin{bmatrix} I \\ P \end{bmatrix} = 0$$

and

$$\mathcal{S}(e_1 \cdots e_d) = \begin{pmatrix} X \\ Y \end{pmatrix},$$

then

$$\bar{P} = YX^{-1}.$$

Remark: We can get other solutions of the Riccati equation by exchanging roots to form different factorizations (i.e., mixing the factors, pushing unstable roots back and forth). Note that if all the roots of the characteristic polynomial are real, there will generally be 2^d solutions.

Remark: We can use the Bass-Roth theorem for periodic cases or even as the starting point for numerical iteration if the system is large.

Lecture 10

Advanced Topics

We consider the correlated noise problem for autonomous single-output linear filtering problems; for details see [41].

Assuming that the matrix Φ is invertible, we consider the following signal process model:

$$x_{n+1} = \Phi x_n + G w_n, \quad x_0 = c$$

$$z_n = h x_n \quad \text{(scalar observations)}.$$

The relevant statistical quantities are

$$l_j = \sum_{i=1}^{j} u_{j-i} u'_{-i}, \quad \text{where } u_i = h\Phi G,$$

$$w_j = \begin{pmatrix} l_1 \\ \vdots \\ l_j \end{pmatrix} \quad j \geq 1, \quad w_0 \equiv 0.$$

Note that the ls in this equation are scalars, as our model has scalar observations. Now, the relevant Riccati equation in the original Kalman form is

$$P_{n+1} = \Phi P_n \Phi' - \Phi P_n h' (h' P_n h)^{\#} h P_n \Phi' + GG'.$$

Define invariant directions as follows: $\alpha \in R^d$ is k-invariant iff

$$\Pi(n,\Gamma)\alpha = \Pi(k,0)\alpha \quad \forall n \geq k, \quad \forall \Gamma \in \mathcal{C}.$$

Note there is always a 1-invariant direction, this is so as $\Phi'^{-1}h'$ is 1-invariant.

Let $e_i = \Phi'^{-i}h'$ be a basis with d^* the number of linear independent e_is. Then we have the following.

Theorem 10.1

$$\sigma \text{ is k-invariant} \Rightarrow \exists f_i \text{ such that } \sigma = \sum_{j=1}^{\min(k,d^*)} f_j e_j.$$

This result is a consequence of the following theorem.

Theorem 10.2 *Suppose $a \leq d^*$. Then e_i is i-invariant ($i = 1,\ldots,a$), $I = Ia$ and $I_j = \text{span}(e_i, \cdots, e_j)$ for $j \leq a$ iff:*

1. $a < d^*, w_a \neq 0, \quad w_{a-1} = 0$

or

2. $a = d^*, \quad w^*_{a-1} = 0.$

This last theorem is called the "step-down lemma," and its proof can be found in [41]. To illustrate what is happening here, consider the following Laplace transform (i.e., spectral density of the output covariance of the signal process model assumed here):

$$\frac{\chi(s^2)}{\Phi(s^2)} \begin{array}{l} 2h \text{ degree} \\ 2d \text{ degree} \end{array} = h(sI - \Phi)^{-1}GG'\left(\frac{1}{s}I - \Phi\right)^{-1}h'.$$

The number of invariant directions is (d^*), and the w_js are the coefficients of the Hurwitz spectral factor of numerator polynomial, which has degree $2(d - d^*)$. The block diagram for the problem in the new basis is shown in Figure 10.1.

1. Invariant Directions

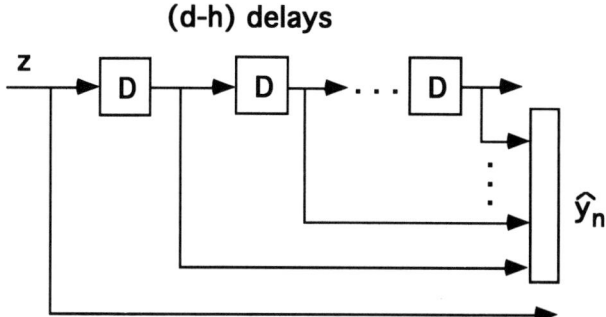

Figure 10.1: Block diagram of the process in the new basis.

This reduces the computational load by realizing the filter in a reduced-size space. Also, this realization of the filter eliminates the stiffness associated with the original Kalman formulation (in that there is now a noise term); however, the covariance of the plant noise, Q, is now negative definite. This can be rectified by adding K_{22} to everything (this will add a new cross-term) since we are interested in the covariance:

$$P_n = \begin{bmatrix} K_{11} & K_{12} \\ K_{12} & M_n + K_{22} \end{bmatrix}.$$

K_{11} is $d^* x d^*$ and only $d - d^*$ estimates are dynamic, with the remainder outputs of a tapped delay line. The theory of the general case of an autonomous model with multiple outputs is considerably more complex; see [29].

1 Invariant Directions

Let's look at the special case of a single-output system with no observation noise:

$$P_{n+1} = \phi(P_n - P_n h(h' P_n h)^\# h' P_n)\phi' + GG' \qquad P_o = \Gamma.$$

Consider

$$h'\phi^{-1}P_{n+1} = u_{-1}G'.$$

$\forall n \geq 0$ and $\forall \Gamma$, this is constant, and the cross correlation between it and any other direction is constant. Therefore, this is an invariant direction. Next, consider

$$h'\phi^{-2}P_{n+2} = \left(h'\phi^{-1}P_{n+1} - h'\phi^{-1}P_n h(h'P_n h)^{\#} h'P_n\right)\phi'$$

$$-u_{-1}u'_o(h'P_n h)^{\#} h'P_n + u_{-2}G'.$$

This will be invariant provided $u_{-1}u'_o = 0$, and the two steps have two invariant directions. This is saying the filter is a tap delay line in k directions, and a feedback system in $d-k$ directions. Hence, proper coordinates for the filter will allow it to be reduced to the form

$$P_n = \begin{bmatrix} S_n & A \\ A' & C \end{bmatrix},$$

where S_n satisfies a lower-dimensional Riccati equation. For further details see [41].

2 Nonlinear Filtering

In this section, the nonlinear filter equations are developed. Assume the equations are autonomous without loss of generality since the time can always be added to the state as an additional variable. Further assume that the state $\mathbf{x}_n \in R^d$ has plant noise $\mathbf{u}_n \in R^r$ and the measurements $\mathbf{z}_n \in R^s$ have measurement noise $\nu_n \in R^s$. Then the model equations are

$$\mathbf{x}_{n+1} = \Phi(\mathbf{x}_n) + \sigma(\mathbf{x}_n)u_n, \quad \mathbf{x}_0 = \mathbf{c}$$

$$\mathbf{z}_n = h(x_n) + \nu_n,$$

2. Nonlinear Filtering

where the $\mathbf{u}_n, \nu_n$, and $\mathbf{c}$ are independent Gaussian random sequences satisfying

$$\mathrm{E}\{\mathbf{u}_n \mathbf{u}'_m\} = Q\delta_{nm}, \quad \mathrm{E}\{\mathbf{cc}'\} = \Gamma \geq 0, \quad \mathrm{E}\{\nu_n \nu'_m\} = R\,\delta_{nm}.$$

We want to find the one-step predictor conditional probability density, denoted as

$$P(\mathbf{x}_{n+1} \mid \mathbf{z}_0, \ldots, \mathbf{z}_n) = P_{n+1}(\mathbf{x}),$$

and the filter conditional density, denoted as

$$P(\mathbf{x}_n \mid \mathbf{z}_0, \ldots, \mathbf{z}_n) = F_n(\mathbf{x}).$$

We want to find equations that sequentially update these densities. We will use Bayes' Rule, "the inverse conditional probability theorem," to derive our results. In this section we often abuse notation, by mixing random variables and their values. Recall that

$$P(\mathbf{z}_0, \mathbf{z}_1, \ldots, \mathbf{z}_n) = \mathrm{E}\{P(\mathbf{z}_0, \mathbf{z}_1, \ldots, \mathbf{z}_n \mid \mathbf{x}_0, \mathbf{x}_1, \ldots, \mathbf{x}_n)\}.$$

Remember that the conditional density is itself a random variable! Given $(\Omega, \mathcal{A}, P)$ and $\mathcal{B}$ a sub σ-field of $\mathcal{A}$, then the following property is called smoothing:

$$\mathrm{E}(\mathrm{E}f|\mathcal{B}) = \mathrm{E}f, \text{ where } f = I_{[\mathbf{z}_0, \mathbf{z}_1, \ldots, \mathbf{z}_n]},$$

where $I_{[\bullet]}$ denotes the indicator function of the set where the sequence $\mathbf{z}_i$ takes on the prescribed values.

The signal process, $\mathbf{x}_n$, is a Markov process with transition density:

$$a(\mathbf{x}_2, \mathbf{x}_1) = P(\mathbf{x}_2 \mid \mathbf{x}_1)$$

with

$$P(\mathbf{x}_2|\mathbf{x}_1) = \frac{1}{(\det(\sigma\sigma'))^{1/2}} \frac{1}{(2\pi)^{r/2}} \exp\left\{-\tfrac{1}{2}\|\mathbf{x}_2 - \Phi(\mathbf{x}_1)\|^2_{[\sigma\sigma'(x_1)]^{-1}}\right\}.$$

Using the Markov property it follows that

$$E\{P(\mathbf{z}_0, \mathbf{z}_1, \ldots, \mathbf{z}_n) \mid \mathbf{x}_0, \mathbf{x}_1, \ldots, \mathbf{x}_n)\} =$$

$$P(\mathbf{z}_0, \ldots, \mathbf{z}_n) = \underbrace{\int\int \cdots \int}_{n+1 \text{ times}} P(\mathbf{z}_0, \ldots, \mathbf{z}_n \mid \mathbf{x}_0, \ldots, \mathbf{x}_n)$$

$$\cdot a(\mathbf{x}_n, \mathbf{x}_{n-1}) \cdots a(\mathbf{x}_1, \mathbf{x}_0) g(\mathbf{x}_0) d\mathbf{x}_0 \cdots d\mathbf{x}_n,$$

where $g(\mathbf{x}_0)$ is the prior density of $\mathbf{x}_0$. Hence,

$$P(\mathbf{z}_0, \ldots, \mathbf{z}_n) \propto \underbrace{\int\int \cdots \int}_{n+1 \text{ times}} \exp - \overbrace{\left\{\tfrac{1}{2}\sum_{i=0}^{n}\|\mathbf{z}_i - h(\mathbf{x}_i)\|^2_{R^{-1}}\right\}}^{H_n}$$

$$\cdot a(\mathbf{x}_n, \mathbf{x}_{n-1}) \cdots a(\mathbf{x}_1, \mathbf{x}_0) g(\mathbf{x}_0) d\mathbf{x}_0 \cdots d\mathbf{x}_n$$

(think of H_n as the energy). Here the integrand is the joint probability of the zs and xs:

$$P(\mathbf{z}_0, \mathbf{x}_0, \ldots, \mathbf{z}_n, \mathbf{x}_n) = d(\underbrace{\mathbf{c}, \mathbf{u}_0 \cdots \mathbf{u}_{n-1}, \mathbf{v}_0 \cdots \mathbf{v}_n}_{\text{all independent}}).$$

To get $P(\mathbf{z}_0, \mathbf{x}_0, \cdots, \mathbf{z}_{n+1}, \mathbf{x}_{n+1})$ just add $a(\mathbf{x}_{n+1}, \mathbf{x}_n)$ to the preceding equation. Normalizing, we obtain

$$P(\mathbf{x}_0, \ldots, \mathbf{x}_n \mid \mathbf{z}^n) =$$

$$\frac{\exp\{-H_n(\mathbf{z}^n, \mathbf{x}^n)\} \prod_{i=1}^{n} a(\mathbf{x}_i, \mathbf{x}_{i-1}) g(\mathbf{x}_0)}{\int_{\mathbf{x}_0} \cdots \int_{\mathbf{x}_n} \exp\{-H_n(\mathbf{z}^n, \mathbf{x}^n)\} \prod_{i=1}^{n} a(\mathbf{x}_i, \mathbf{x}_{i-1}) g(\mathbf{x}_0) d\mathbf{x}_0 \cdots d\mathbf{x}_n}$$

2. Nonlinear Filtering

and

$$P(\mathbf{x}_0, \ldots, \mathbf{x}_{n+1} \mid \mathbf{z}^n) =$$

$$\frac{\exp\{-H_n(\mathbf{z}^n, \mathbf{x}^n)\} \prod_{i=1}^{n+1} a(\mathbf{x}_i, \mathbf{x}_{i-1}) g(\mathbf{x}_0)}{\int_{\mathbf{x}_0} \cdots \int_{\mathbf{x}_{n+1}} \exp\{-H_n(\mathbf{z}^n, \mathbf{x}^n)\} \prod_{i=1}^{n+1} a(\mathbf{x}_i, \mathbf{x}_{i-1}) g(\mathbf{x}_0) d\mathbf{x}_0 \cdots d\mathbf{x}_{n+1}},$$

where conditioning on $\mathbf{z}^n$ indicates we are given $\mathbf{z}_0, \mathbf{z}_1, \ldots, \mathbf{z}_n$. We did this by using Bayes' Rule (or equivalently by noting the independence of the integrand and the relation to the $\mathbf{z}$s and $\mathbf{x}$s).

Now, integrate the appropriate $\mathbf{x}'_i s$ in the above equations to obtain $P(\mathbf{x}_n \mid \mathbf{z}^n)$ and $P(\mathbf{x}_{n+1} \mid \mathbf{z}^n)$, and write the above in the nicer form:

$$F_n(\mathbf{x}) = \frac{\mu(\mathbf{x}_n) \underbrace{\int \cdots \int}_{n \text{ times}} \exp\{-H_n(\mathbf{z}^n, \mathbf{x}^n)\} P(\mathbf{x}_0, \ldots, \mathbf{x}_{n-1} \mid \mathbf{x}_n) d\mathbf{x}_0 \cdots d\mathbf{x}_{n-1}}{\underbrace{\int \cdots \int}_{n+1 \text{ times}} \exp\{-H_n(\mathbf{z}^n, \mathbf{x}^n)\} P(\mathbf{x}_0, \ldots, \mathbf{x}_{n-1} \mid \mathbf{x}_n) \mu(\mathbf{x}_n) d\mathbf{x}_0 \cdots d\mathbf{x}_n},$$

where $\mu_n(\mathbf{x}_n) = P(\mathbf{x}_n)$, the unconditioned density of $\mathbf{x}_n$. Now also,

$$P_{n+1}(\mathbf{x}) = \frac{\underbrace{\int \cdots \int}_{n+1 \text{ times}} \exp\{-H_n(\mathbf{z}^n, \mathbf{x}^n)\} P(\mathbf{x}_0, \ldots, \mathbf{x}_n \mid \mathbf{x}_{n+1}) d\mathbf{x}_0 \cdots d\mathbf{x}_n \, \mu(\mathbf{x}_{n+1})}{\underbrace{\int \cdots \int}_{n+2 \text{ times}} \exp\{-H_n(\mathbf{z}^n, \mathbf{x}^n)\} P(\mathbf{x}_0, \ldots, \mathbf{x}_n \mid \mathbf{x}_{n+1}) \mu(\mathbf{x}_{n+1}) d\mathbf{x}_0 \cdots d\mathbf{x}_{n+1}}.$$

If the index of this relation is changed from $n = 1$ to n, then the numerator and denominator integrands of $P_n(\mathbf{x})$ and $F_n(\mathbf{x})$ are the same except for the term $\exp(-H_n)$. Explicitly,

numerator integrand of $F_n(\mathbf{x}) =$

$$\exp\left\{-\tfrac{1}{2}\|\mathbf{z}_n - h(\mathbf{x}_n)\|^2_{R^{-1}}\right\} \times \text{numerator integrand of } P_n(\mathbf{x}).$$

This leads to the following lemma.

Lemma 10.1 *The filter density at the nth step may be written in terms of the predictor density from the previous step as*

$$F_n(\mathbf{x}) = \frac{\exp\left\{-\tfrac{1}{2}\|\mathbf{z}_n - h(\mathbf{x}_n)\|^2_{R^{-1}}\right\} P_n(\mathbf{x})}{N_n},$$

where N_n normalizes the density so that it integrates to unity. If the densities are non-Gaussian, then replace the exponential term in the above formula with the correct density.

Recall the analogous equation in the linear case as

$$\hat{\mathbf{x}}_{n|n} = K_n(\mathbf{z}_n - H\,\hat{\mathbf{x}}_{n|n-1}).$$

Lemma 10.2 *The predictor density is a generalized convolution of the filter density:*

$$P_{n+1}(\mathbf{x}) = (S * F_n)(\mathbf{x}).$$

To see this, take the numerator of F_n:

$$\int a(\mathbf{x}_{n+1}, \mathbf{x})\,[\text{ numerator of } F_n(\mathbf{x})]\,d\mathbf{x}_{n+1} =$$

$$\underbrace{\int \cdots \int}_{n+1 \text{ times}} \exp\{-H_n(\mathbf{z}^n, \mathbf{x}^n)\} P(\mathbf{x}_0, \ldots, \mathbf{x}_n)\, P(\mathbf{x}_{n+1} \mid \mathbf{x}_n)\, d\mathbf{x}_0 \cdots d\mathbf{x}_n.$$

2. Nonlinear Filtering

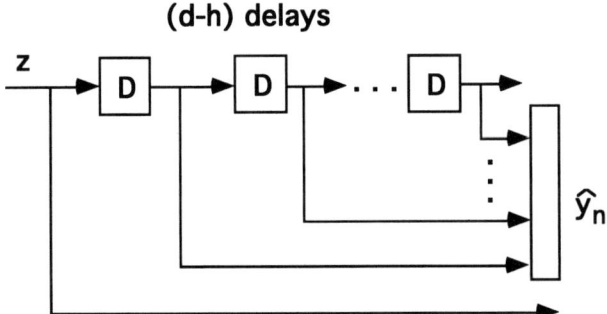

Figure 10.2: System diagram for the nonlinear filter.

But by the Markov property,

$$P(\mathbf{x}_0, ..., \mathbf{x}_n) P(\mathbf{x}_{n+1} \mid \mathbf{x}_n) = P(\mathbf{x}_0, ..., \mathbf{x}_n) P(\mathbf{x}_{n+1} \mid \mathbf{x}_n, ..., \mathbf{x}_0)$$

$$= P(\mathbf{x}_0, ..., \mathbf{x}_{n+1})$$

and hence,

$$\int a(\mathbf{x}_{n+1}, \mathbf{x}) \, [\text{ numerator of } F_n(\mathbf{x})] \, d\mathbf{x}_{n+1} = \text{ numerator of } P_{n+1}(\mathbf{x}).$$

This is the analog of

$$\hat{\mathbf{x}}_{n+1|n} = \Phi \hat{\mathbf{x}}_{n|n}$$

in the linear case. Note that

$$S(\mathbf{x}, \mathbf{y}) = a(\mathbf{y}, \mathbf{x}) \propto \frac{1}{\det(\sigma\sigma')^{\frac{1}{2}}} \exp\left\{-\frac{1}{2} \|\mathbf{y} - \Phi(\mathbf{x})\|^2_{[\sigma\sigma']^{-1}}\right\}$$

and thus this is not a true convolution since it depends on $\mathbf{y} - \Phi(\mathbf{x})$ rather than on $\mathbf{y} - \mathbf{x}$. The system diagram for this filter is shown in Figure 10.2.

In Figure 10.2, the filter density is $F_n(\mathbf{x}) = f(\mathbf{x}_1, ..., \mathbf{x}_n, \mathbf{z}_0, ..., \mathbf{z}_n)$

and thus at stage n F_n depends on $(d+s)n$ variables, but we can represent this for known z, or as a function of d variables. Bellman called this situation, the need to know a function in place of a vector, the "curse of dimensionality."

Note that if we perform a Fourier transform of the filter equations, it changes $S*$ into a multiply but the data equation into a convolution so that the computational load remains the same. Notice the similarity to a testing problem,

$$H_0 \quad z_n = h(x_n) + \nu_n \qquad n \in T$$

$$H_1 \quad z_n = \nu_n.$$

When N_n, the normalizing constant, is small, we reject the null hypothesis, i.e., H_0. This was one of the results of Tyrone Duncan's thesis; see the references in [42]. The optimal test—the likelihood ratio test—is related to the N_n in our data-update equation. This leads to the statement that information theory is a subdiscipline of nonlinear filtering. Testing problems are relevant to surveillance and tracking; i.e., they are important. For references and additional details, see [42].

Lecture 11

Applications

1 Historical Applications

Recall from the previous lecture that the conditional density is given by

$$P^{x_n|z_0\ldots z_n}(\xi) \triangleq F_n(\xi) = \frac{D_n(\xi)P_n(\xi)}{N_n},$$

where D_n is

$$D_n(\xi) = \nu_n(z_n - h(\xi)),$$

ν_n is the density of the nth observation noise, and

$$N_n = \int \cdots \int D_n(\xi) f_n(\xi)\, d\xi$$

is the normalizing factor.

Now, using the relations between F_n and P_n and P_n and F_{n-1}, it follows that

$$P^{x_n|z_0\ldots z_{n-1}}(\mu) \triangleq F_n(\mu) = \underbrace{\int \cdots \int}_{r \text{ times}} S_n(\mu, \lambda) F_{n-1}(\lambda)\, d\lambda$$

117

where

$$S_n(\mu, \lambda) = g\left[\sigma^{\#}(\lambda)\left(\mu - \phi(\lambda)\right)\right],$$

g is the nth plant noise density, and

$$P_o(\xi) \equiv \gamma(\xi)$$

with γ the density of c.

Recall our model:

$$\mathbf{x}_{n+1} = \phi(\mathbf{x}_n) + \sigma(\mathbf{x}_n)\mathbf{u}_n$$

$$\mathbf{x}_0 = \mathbf{c}$$

$$\mathbf{z}_n = h(\mathbf{x}_n) + \nu_n,$$

where all the processes are multidimensional.

<u>Remark 1</u>: Finding $\sigma^{\#}$ can be a bit of a problem.

<u>Remark 2</u>: When the noise does not enter some of the state variables, S_n collapses to a delta function for those variables.

We want to consider several problems.

1.1 Problem 1. Cubic Sensor Problem (d=1)

$$x_{n+1} = \alpha x_n + u_n \qquad \mathrm{E}u_n^2 = 1$$

$$z_n = x_n^3 + \nu_n \qquad \mathrm{E}\nu_n^2 = r,$$

where u_n is white noise.

1. Historical Applications

We want this problem to be observable; it has hopes of being observable because if $r = 0$ (no noise) we just take the cube root. In order to understand the observability question for nonlinear stochastic systems it is useful to consider the continuous-time version of our problem. These notes are not aimed at giving a complete picture of this complex subject, which is treated in detail in [42], so we include only a summary of the main features of the continuous-time problem.

<u>The Analogous Continuous-Time Problem</u>: Let $r \to \frac{r}{\Delta}$, $q \to \Delta q$, and $\alpha \to (1 + \Delta \gamma)$ as $\Delta \downarrow 0$. Then we obtain

$$dx = \gamma x \, dt + d\beta$$

$$dz = x^3 \, dt + d\nu.$$

The Markov process associated with this continuous-time problem has a transition density that satisfies partial differential equations called the forward and backward equations. Solutions x_t (interpreted by the Ito calculus) have continuous paths and a transition probability

$$p(t, x, y) dy = P(x_t \in dy | x_o = x),$$

which is the cumulative probability that x_t lies between y and $y \pm dy$. These transition densities satisfy the Chapman-Kolmogorov equation:

$$p(t + s, z, y) = \int_{-\infty}^{\infty} p(s, z, x) p(t, x, y) \, dx.$$

In 1933, Kolmogorov proved that the equation

$$dx = \gamma x \, dt + d\beta$$

has an abstract meaning as follows, if

$$E\{(x_{t+h} - x_t) | x_t\} = \gamma x_t h + o(h)$$

$$\mathrm{E}\{(x_{t+h} - x_t)^2 | x_t\} = h + o(h).$$

These are the Lindeberg conditions, which follow from the central limit theorem. More generally, if

$$dx = f\,dt + \sigma\,d\beta,$$

then p, the transition probability, satisfies

$$\frac{\partial p}{\partial t} = f\frac{\partial p}{\partial x} + \frac{1}{2}\sigma^2\frac{\partial^2 p}{\partial x^2} = Ap,$$

which is the backward equation. A is the backward operator. There is an analogous forward equation

$$-\frac{\partial p}{\partial t} = -\frac{\partial}{\partial x}(fp) + \frac{1}{2}\frac{\partial^2}{\partial x^2}(\sigma^2 p) = \tilde{A}p,$$

which is the Fokker-Planck equation. Kolmogorov proved this by using the Chapman-Kolmorgorov equation:

$$P(x_{t+h} \in dy | x_o) = \int P(x_{t+h} \in dy | x = x_t) P(x_t = x | x_o)\,dx$$

and then taking the limit as $h \to 0$ to get $\frac{\partial p}{\partial t}$. See Kolmogorov's paper [43] for the details.

Integrating the Fokker-Planck equation by parts, we get

$$\mathrm{E}\{h(x_t)|x_o = x\} = h(x) + \int_0^t \mathrm{E}\left\{\tilde{A}h(x_s)|x_o = x\right\}ds,$$

where we do this to get the average of a function (Dynkin's formula; see [1]). Observability is then that the rank of the sensor orbit is full, i.e., equal to the signal process state dimension:

$$(h, \tilde{A}h, \tilde{A}^2 h, \ldots, \tilde{A}^n h, \ldots).$$

1. Historical Applications

For our cubic sensor, we have

$$(x^3, \gamma x^3 + \tfrac{1}{2}x, \ldots),$$

where the sensor orbit has degree 2. By introducing sensor orbit coordinates, we can bound the filter performance. Sensor orbit coordinates for the cubic sensor problem in the continuous-time case can be introduced as follows. Recall the original problem as

$$\begin{aligned} dx &= \gamma x dt + d\beta \\ dz &= x^3 dt + d\nu. \end{aligned}$$

Now introducing sensor orbit coordinates, let $y_1 = x$ and $y_2 = x^3$. Then the system becomes

$$\begin{aligned} dy_1 &= \gamma y_1 dt + d\beta \\ dy_2 &= (3\gamma y_2 + 3y_1) dt + 3x^2 d\beta \\ dz_1 &= y_2 dt + d\nu. \end{aligned}$$

The x^2 factor in the equation for dy_2 can be replaced by its statistical linearization, $\frac{q}{2\beta}$. The resulting system is linear, so we can find the solution to the Riccati equation. Note that p_{11} is an upper bound on the original problem.

1.2 Numerical Realization

Now let's make a nonlinear filter for our cubic sensor problem. At step n, we know p_{n-1}, μ_{n-1} (the conditional mean of p_{n-1}), and σ_{n-1} (the standard deviation of the second central moment). To find p_n, we represent the density in terms of a set of point masses:

$$p_n(x) = \sum_{m=-N}^{N} \xi_m^n \delta(x - x_m^n),$$

where

$$x_m^n = \mu_{n-1} + \frac{k\sigma_{n-1}(m)}{N}.$$

This is the "floating grid," in which the mesh gets finer as you get better estimates. We want to establish a recursion relation for the ξ_m^n.

$$p_{n+1}(\mu) = \int e^{\frac{-(\mu-\alpha\lambda)^2}{2}} \frac{1}{N_n} e^{\frac{-(z_n-h(\lambda))^2}{2r}} p_n(\lambda)\, d\lambda,$$

where N_n is a normalizing constant. Thus

$$\sum_{m=-N}^{N} \xi_m^{n+1} \delta(\mu - x_m^{n+1})$$

$$= \int e^{\frac{-(\mu-\alpha\lambda)^2}{2}} e^{\frac{-(z_n-h(\lambda))^2}{2r}} \sum \frac{1}{N_n} \xi_m^n \delta(\lambda - x_m^n)\, d\lambda.$$

Hit this with a function that is 1 around x_m^{n+1} and zero elsewhere, and we obtain

$$\xi_m^{n+1} = \frac{1}{N_n} \sum_{k=-N}^{N} e^{\frac{-(x_k^{n+1}-\alpha x_k^n)^2}{2}} e^{\frac{-(z_n-h(x_m^n))^2}{2r}} \xi_k^n.$$

Note that if all the ξ_os are positive, by induction the ξs are positive all the time. Note that this problem does not have a "fading memory" for previous measurements. Let

$$\hat{x}_n = \frac{1}{N_n} \sum x_m^n \xi_m^n.$$

If we have a large z_n, simulation results (see, for example, [44]) show the new estimate is

$$\hat{x}_n = (z_n)^{\frac{1}{3}}$$

and all the old data are forgotten.

1. Historical Applications

Another way of handling this problem is to expand x_n^3 about $\hat{x}_n$:

$$x_n^3 = \hat{x}_{n|n-1}^3 + 3\hat{x}_{n|n-1}^2 x_n + \cdots .$$

We then subtract $\hat{x}_{n|n-1}^3$ and define a new sensor

$$z_n^* = 3\hat{x}_{n|n-1}^2 x_n + \nu_n$$

that is quasi-linear. The update equation would then be

$$\hat{x}_{n+1|n} = \alpha \hat{x}_{n|n-1} + \frac{3 p_n \hat{x}_n^2}{q p_n \hat{x}_n^4 + r}(z_n^* - 3\hat{x}_{n|n-1}^3),$$

and the Riccati equation would be

$$p_{n+1} = \left(\frac{\alpha^2}{p_n^{-1} + \frac{q}{r}\hat{x}_{n|n-1}^4} \right) + q \qquad p_0 = \Gamma.$$

The new filter and Riccati equations are coupled through the sensor. We can study this system with Monte Carlo techniques. If Γ is small, we get convergence for almost any observation noise r (with no plant noise), but if Γ is large, the system diverges. In the multidimensional case, similar methods can be used.

1.3   Problem 2. Passive Receiver

This problem arises in two dimensions for the Navy, and in six dimensions for the Air Force (AWACS). An observer moves on a circle and observes a target (assumed fixed). The line of sight makes an angle α with the x^1 axis (see Figure 11.1). The model is described by

$$x_{n+1}^1 = \alpha x_n^1 + \beta x_n^2 + u_n^1, \quad \mathrm{E}(u_n^1)^2 = q_1$$

$$x_{n+1}^2 = \gamma x_n^1 + \delta x_n^2 + u_n^2, \quad \mathrm{E}(u_n^2)^2 = q_2$$

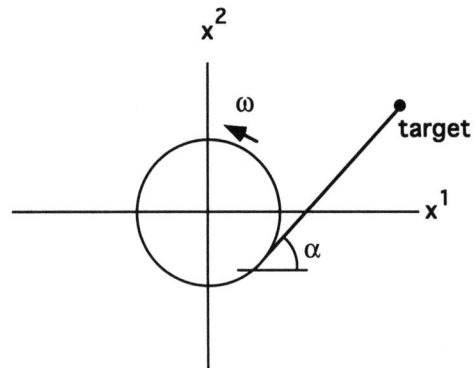

Figure 11.1: Two-dimensional passive receiver geometry.

$$z_n = \tan^{-1}\left(\frac{x_n^2 - \sin n\Delta}{x_n^1 - \cos n\Delta}\right) + \nu_n, \quad E\nu_n^2 = r.$$

The point mass method works well on this problem, while the extended linear and second-order filters do not; the nonlinear filter for this problem does not in general have a closed-form solution. If we have no plant noise on one state variable, the generalized convolution equation simplifies as one integration goes away.

For problem 2 (passive receiver), we have

$$P_{n+1}(\mu_1, \mu_2) = \iint \frac{1}{2\pi(q_1 q_2)^{1/2}} e^{-\frac{(\mu_1 - \alpha x - \beta y)^2}{2q_1}} e^{-\frac{(\mu_2 - \gamma x - \delta y)^2}{2q_2}} F_n(x, y)\, dx\, dy.$$

Now, as $q_1 \to 0$, the two-fold convolution reduces to one-fold:

$$P_{n+1}(\mu_1, \mu_2) = \int \frac{1}{\sqrt{2\pi q_2}} e^{-\frac{(\mu_2 - \gamma x - \frac{1}{\beta}(\mu_1 - \alpha x))^2}{2q_2}} F_n\left(x, \frac{\mu_1 - \alpha x}{\beta}\right) dx.$$

<u>Comment</u>: If we have d dimensions and m point masses in each dimension, the vector length is m^d, and to do the convolution we need to multiply

1. Historical Applications

an $m^d \times m^d$ matrix by an m^d vector, which requires m^{3d} operations. If $d = 6$ and $m = 20$, the number of operations exceeds 10^{23}. Because of all the dot-product operations, the problem can be solved much faster on a parallel machine.

Lecture 12

Phase Tracking

1 The Phase Lock Loop

Phase tracking is an important problem. It is a filtering problem consisting of an observation process,

$$z(t) = A\cos(\omega_0 t + \phi) + n(t),$$

and a signal process, ϕ, which for didactic purposes we will initially assume is a Brownian motion process:

$$d\phi = d\beta, \quad \phi_0 = c.$$

The phase, ϕ, is of interest; ω_0 is the carrier frequency, and $n(t)$ is the noise. We want to seperate the very large term, $\omega_0 t$, from the term ϕ. This is referred to as heterodyning down to base band, and can be accomplished by multiplying the observation process by sinusoids and low-pass filtering the result. Schematically, these operations are

(1) $\quad z(t) \xrightarrow{\quad} \otimes \xrightarrow{2\cos\omega_0 t} [A\cos\phi + A\cos(2\omega_0 t + \phi) + 2n(t)\cos\omega_0 t]$

(2) $\quad z(t) \xrightarrow{\quad} \otimes \xrightarrow{2\sin\omega_0 t} [A\sin\phi + A\sin(2\omega_0 t + \phi) + 2n(t)\sin\omega_0 t]$

$$\text{(1)} \quad \longrightarrow< \text{lowpass} =\!\!\longrightarrow A\cos\phi + 2n\sin\delta$$

$$\text{(2)} \quad \longrightarrow< \text{lowpass} =\!\!\longrightarrow A\sin\phi + 2n\cos\delta$$

where $\delta = \omega_0 t$.

We assume the noise $n(t)$ is ergodic; then $n(t)\cos\delta$ and $n(t)\sin\delta$ are uncorrelated, and our observation becomes

$$dz_1 = \cos\phi\, dt + dv_1$$

$$dz_2 = \sin\phi\, dt + dv_2$$

with v_1 and v_2 independent Brownian motions of variance r.

In order to estimate the phase, ϕ, classically a feedback loop

$$d\hat{\phi} = k(dz_2\cos\hat{\phi} - dz_1\sin\hat{\phi}) \approx k(\sin(\phi - \hat{\phi}) + \text{noise}) = kdI$$

is used. This is called a phase lock loop, PLL, and is an extended Kalman-Bucy filter. For small phase error $\sin(\phi - \hat{\phi}) \approx \phi - \hat{\phi}$, so

$$kdI \approx k(\phi - \hat{\phi} + dv)$$

and linear theory can be applied. Let $\tilde{\phi} = \phi - \hat{\phi}$ be the phase error. Then the error estimate equation becomes

$$d\tilde{\phi} = -k\tilde{\phi}dt - kdv + d\beta.$$

Now the error variance is the solution of

$$\dot{p} = -2kp + rk^2 + q$$

and minimizing the right-hand side of this equation for the gain k yields

1. The Phase Lock Loop

the condition

$$2rk - 2p = 0,$$

which in turn gives rise to the Riccati equation for the optimal filter for the linearized problem, with optimal steady-state gain $k = \frac{p}{r} = (\frac{q}{r})^{\frac{1}{2}}$. The optimal steady-state filter is

$$d\hat{\phi} = \sqrt{\frac{q}{r}}\,(dz_2 \cos\hat{\phi} - dz_1 \sin\hat{\phi}).$$

Of course, the Wiener theory can also be applied to the linearized problem as in Lecture 4 and will result in the same design. To summarize,

$$\begin{aligned} k &= p(t)/r && \text{is used for transient optimal gain} \\ k &= \sqrt{q/r} && \text{is steady-state gain optimal gain} \\ p &= \sqrt{qr} && \text{is the steady-state error variance} \end{aligned}$$

for the small phase error situation. The exact error can be found using the Fokker-Planck equation since

$$d\hat{\phi} = \alpha(\sin\phi - \hat{\phi})\,dt + d\nu$$

$$d\tilde{\phi} = d\beta - \alpha d\nu - \alpha \sin\tilde{\phi}.$$

It follows that the density function of the phase error satisfies the Fokker-Planck equation:

$$\frac{\partial p}{\partial t} = -\alpha \sin x \frac{\partial p}{\partial x} + \frac{1}{2}(q + \alpha^2 r)\frac{\partial^2 p}{\partial x^2},$$

or the steady-state density satisfies

$$\frac{\partial p}{\partial t} = \frac{\partial}{\partial x}(\alpha \sin xp) + q\frac{\partial^2 p}{\partial x^2} = 0.$$

A solution is

$$\frac{\partial p}{\partial x} = -\frac{\alpha}{q}\sin xp$$

or

$$p(x) = \frac{e^{-\frac{1}{\sqrt{qr}}\cos x}}{I_0\left(\frac{1}{\sqrt{qr}}\right)}$$

where I_0 is a Bessel function of the first kind. For $\sqrt{qr}$ small, p is Gaussian, and for $\sqrt{qr}$ large, p is uniform.

2 Phase Demodulation

The discrete nonlinear filter for a more general signal model phase demodulation problem will now be discussed. The signal and observation models are

$$dx^1 = x^2\, dt$$

$$dx^2 = d\beta$$

$$dz_1 = \cos x^1\, dt + dv_1$$

$$dz_2 = \sin x^1\, dt + dv_2.$$

The optimal filter is determined by the one step predictor density, which satisfies

$$P_{n+1}(x,y) =$$

$$\frac{1}{N_n}\int_{-\pi/\Delta}^{\pi/\Delta} e^{\frac{-(u-y)^2}{q}}\, e^{z_n^1\cos(x-\Delta u)+z_n^2\sin(x-\Delta u)}\, P_n(x-\Delta u, u)\, du$$

Notice since we observe and wish to estimate only modulo 2π, the distribution lives on the torus $(-\pi,\pi)\times(-\pi/\Delta,\pi/\Delta)$, after the predictor density is made periodic in phase and phase rate. The details of the derivation of this equation can be found in [45]. The signal model has been discretized with

2. Phase Demodulation

time step Δ and the density represented by point masses. How should we estimate the phase angle? We want to estimate in direction of "heaviest" concentration of density on the boundary of the disk.

Mathematically, this estimate is the x^*, which minimizes the cyclic loss;

$$\min_{x^*} \mathrm{E}\{1 - \cos(x_{n+1} - x^*) | z_0, \ldots, z_n\} =$$

$$\int_{-\pi}^{\pi} [1 - \cos(y - x^*)] \, P_{n+1}(y) \, dy$$

and consequently must satisfy

$$\int_{-\pi}^{\pi} \sin(y - x^*) \, P_{n+1}(y) \, dy = 0.$$

This last equation is easy to solve; the solution is

$$x^* = \tan^{-1} \frac{\widehat{\sin}_{n+1}}{\widehat{\cos}_{n+1}},$$

where for an arbitrary function g, $\widehat{g_n} = \int g(x) P_n(x) dx$; hence, the notation $\widehat{}$ indicates conditioned expectation. The density-update equation has been synthesized on large-scale computers and extensive Monte Carlo error analyses have been performed; see [45] for details.

Lecture 13

Device Synthesis

1 Device Synthesis for Nonlinear Filtering

Several approaches for realizing the nonlinear filter are described in this lecture. These approaches develop new and unique methods for dealing with the required "convolution" mentioned in earlier lectures:

$$P_{n+1} = S * F_n$$

$$F_n = \frac{D_n \cdot P_n}{N_n},$$

where the system block diagram for this filter is shown in Figure 13.1. See [46], Chapter 4 for theoretical details and references to the various synthesis techniques surveyed here.

1.1 Hybrid Computing in Nonlinear Filtering

The thesis by Don Miller (see [47]) in the early 1970s set the theoretical foundation for parallel implementation of the one-dimensional nonlinear filter using hybrid techniques. His research was sponsored by the National Science Foundation for the nonlinear filter implementation in a hybrid (analog-digital) hardware experiment. This research was carried out at the analog computation laboratory of the Universidad Politecnico de Barcelona; see [48]. Since the convolution required by the nonlinear filter is very time-consuming on a digital computer, the filter was divided into analog

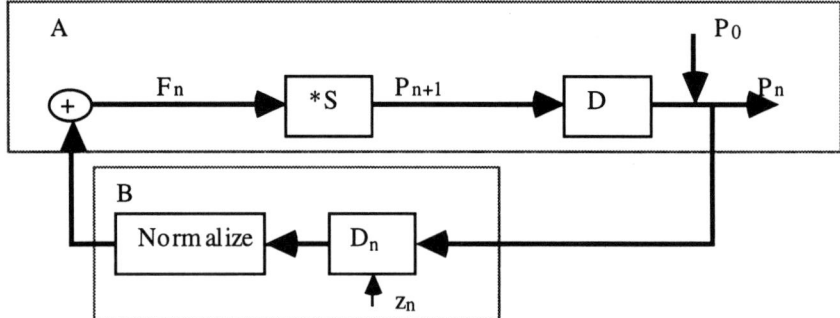

Figure 13.1: Block diagram of the nonlinear filter showing segmentation A for convolution and B for data processing.

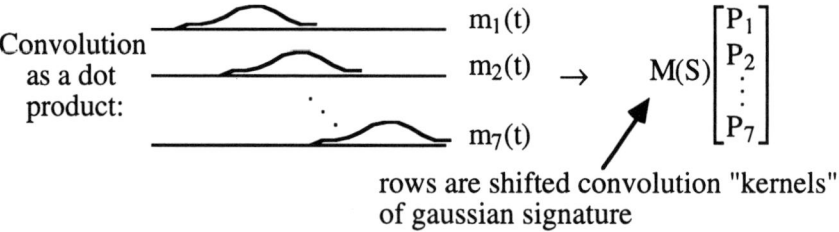

Figure 13.2: Schematic of convolution as a dot product.

and digital segments as indicated in Figure 13.1. The convolution can be thought of as a dot product of a given function, F_n, with appropriately delayed realizations of the kernel S; here we identify space with time. We synthesized the convolution with seven products, each as shown schematically in Figure 13.2. Speed is gained by performing these dot products in parallel using high-speed multipliers. All seven values of the new P are obtained simultaneously; the choice of carrying only seven values of the conditional density was dictated by the number of integrators available.

The problem remaining is how to make the Gaussian density for the

1. Device Synthesis for Nonlinear Filtering

kernels in this implementation. We want

$$f(t) = \begin{cases} \exp\left[(t-\tau)^2/2\right] & \text{for } t \geq \tau - 3 \\ 0 & \text{for } t < \tau - 3 \end{cases}$$

with

$$\dot{f}(t) = -(t-\tau)f(t),$$

where the initial conditions must be fed to the integrator at the instant $t = \tau - 3$. Hence, the generators and multiply are performed in series as shown in Figure 13.3. The tasks of normalization, new data incorporation, $P_n \to F_n$, estimate generation, and gridding are performed digitally. This gives accuracy equivalent to a much larger digital simulation. Note that if we had used a fast Fourier transform simulation of the convolution, then it would not in general be accurate to more than one percent or so.

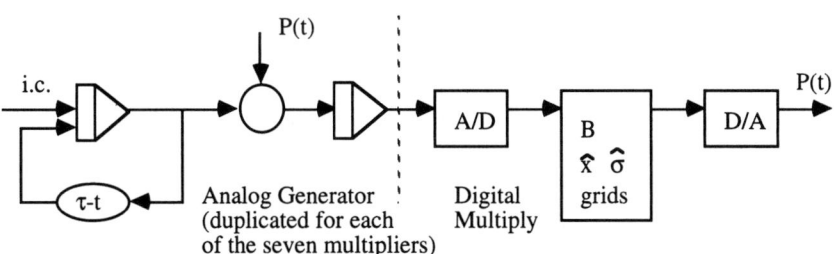

Figure 13.3: Generation of new Ps with hybrid computing hardware.

1.2 Optical Techniques for Nonlinear Filter Convolution

Another idea was to form the necessary convolution based on optics; see [49]. The Air Force supported Bucy, Steier, Mallinckrodt, and Stevens in the period from 1978 to 1981 to extend the above hybrid experiments to use incoherent optical hardware for the convolution. In this technique, a

16 × 16 grid of light-emitting diodes are driven by the analog signals to produce intensities proportional to $f(t)$. These light signals are collected through a lens and passed through a plate frosted with a Gaussian coating. This frosted plate performs the function of the convolution S. Then the light falls on a collector plate made up of many photoelectric detectors, which convert the light signals back to electrical signals. These electrical signals are passed through adders and then on to the feedback loop. This optical hardware is depicted schematically in Figure 13.4.

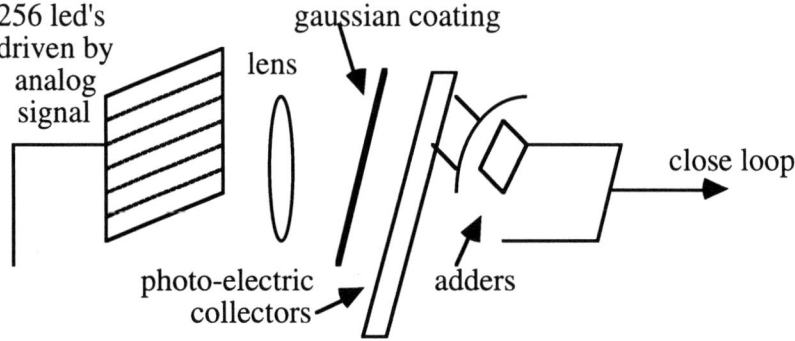

Figure 13.4: Schematic for incoherent optical convolution.

In addition to this optical technique, there is Mallinckrodt's TV idea, which again coded space into time; in this case, R^2 was generalized to R^n (see [50]). With this method, the convolution is performed by conceptually coding time into space by analogy to scans on a TV screen. These 1-D periodic functions of time produce $f * g$. Call these functions $P_1(0, T)$ and say they denote the natural convolution given by

$$f * g(t) = \int_0^T f(t-s)g(s)ds.$$

Then Mallinckrodt wanted to look at functions like $P_2([0,T] \times [0,1])$, which

1. Device Synthesis for Nonlinear Filtering

denote the natural convolution given by

$$F * G(t, s) = \int_0^1 \int_0^1 F(t - \lambda, s - \mu)G(\lambda, \mu)d\lambda d\mu.$$

Then call $t, s \to x, y$ of the TV screen; the map $M(t)$ becomes

$$x(t) = \alpha t \bmod 1$$

$$y(t) = \beta t \bmod 1,$$

where if α/β is rational, then the sweep lines are stable. The map

$$M : [0, T] \to [0, 1] * [0, 1]$$

forms a group, and in fact M is "almost" a homomorphism; i.e.,

$$M(f * g(t)) \approx Mf * Mg.$$

This relation is shown in [51]. We can then define M so that the analog segment of the nonlinear filter shown in Figure 13.1 can be realized as a series of scalar convolutions as depicted in Figure 13.5.

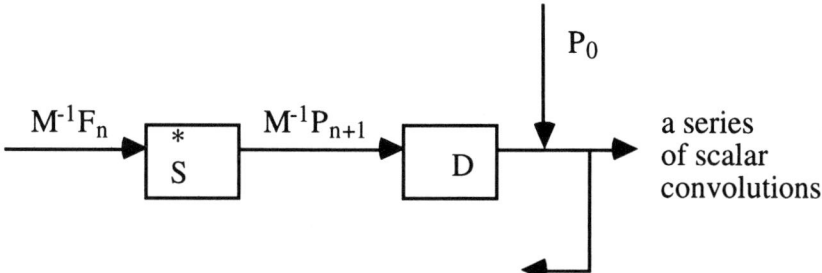

Figure 13.5: Schematic of a single scalar convolution for Mallinckrodt's TV idea.

1.3 Acoustic Techniques for Nonlinear Filter Convolution

Another idea for reducing costs of the nonlinear filter through hardware using surface acoustic wave convolution was proposed and investigated initially with Professor Eugene Dieulesaint and later with Dr. Herve Gautier of Thomson CSF in Valbonne, France, ca. 1983; see [52]. In this technique the filter density time signal, $f(t)$, using the previously described space-into-time coding, is fed through a piezoelectric crystal whose surface is coated with up to 2000 tiny "fingers" of metal. Pure crystals have size limitations, which imply micron interfinger distances if 2000 fingers are required; this represents the state of the art as of 1984. These metal fingers affect the surface waves on the crystal, and in effect code S for the convolution. The output signal at the other end of the crystal is then the time signal corresponding to the new P. This device is shown schematically in Figure 13.6.

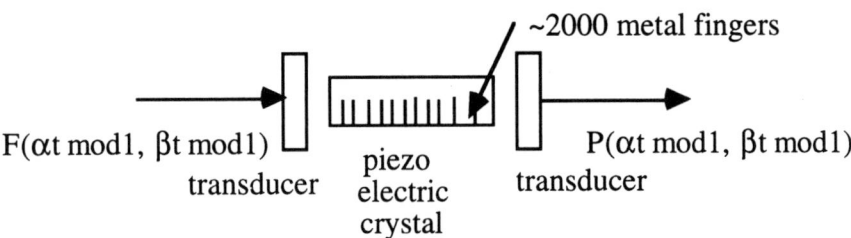

Figure 13.6: Schematic of the surface acoustic wave convolution device.

1.4 Digital Developments in Nonlinear Filtering

Basic advances in the speed of digital computers in the 1970s and 1980s made the accurate numerical computation of the convolution feasible. The most promising machines, which were tested with various nonlinear filtering problems, are listed in Table 13.1. Of these computers, the AP120b was used to produce two- and three-dimensional phase lock loop optimal

1. Device Synthesis for Nonlinear Filtering 139

filters while being hosted by a PDP1155 computer. This configuration was extremely cost-effective; see [53] for a cost comparison.

Table 13.1: Digital computers used to realize nonlinear filters.

Digital Computer Development		
Computer	Facility	Year(s)
Burroughs	Air Force Academy	1969
CDC 6600	Kirkland AFWSL/Eglin AFB	< 1976
Iliac	NASA Moffet Field	1974
CDC 7600	Aerospace Corp.	1970s
Star	NASA Langley	1975
AP120b	USC	
Cray1s	Kirkland	1976
Cray XMP-48	Cray Research	1980s
Cray II	Cray Research	

At Kirkland Air Force Base, Bucy and Senne produced movies of the evolution of the probability distribution over time (some for multiple targets) using the CDC 6600.

Two basic notions in digital computing today are massive parallelism and pipeline architectures. The Cray1s is a pipeline machine that is designed so that the speed of light is the limit of computing speed. This is accomplished by keeping wiring connections between components short and compact placement of components, which in turn results in critical cooling needs. The Cray1s comes with its own cooling hardware under the benches surrounding the central cabinet. In the future, Cray will add more CPUs but retain the pipeline architecture (maybe 16-48 CPUs with shared memory) to try to obtain up to 100 gigaflops (a gigaflop is 1000 million floating-point operations per second). Details of coding the nonlinear filter convolution on the Cray XMP-48 and Cray II can be found in [54].

In spite of the increased usefulness of faster digital machines for nonlinear filtering, the analog method is probably the most promising in terms of future applications (especially the surface acoustic wave devices).

2 Radar Filtering Application

Consider the problem of tracking r targets with multiple looks of a phased-array radar system as depicted in Figure 13.7. The α_is in Figure 13.7 are related by the array spacing, d, and the wavelength, λ, to a new variable, θ, as $\theta = 2d/\lambda \sin \alpha$. There is a complex n-vector of outputs from the array, z:

$$z^j = \sum_{i=1}^{r} a_i^j d(\theta_i) + v^j,$$

where $E(vv^*)$ is diagonal and can be normalized to identity. $E(a_i^k a_j^{l*}) = \rho_i \delta_{ij} \delta_{kl}$ is Gaussian

If the array is equally spaced,

$$d(\theta) = \begin{pmatrix} e^{i\theta} \\ e^{2i\theta} \\ \vdots \\ e^{ni\theta} \end{pmatrix}.$$

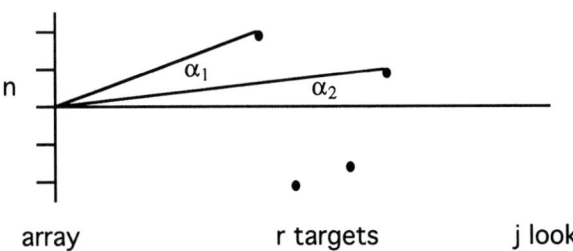

Figure 13.7: Phased array radar filtering problem.

The problem now is to estimate the θ_is. The amazing thing is that we can find the θ_is without finding the a_is. A good reference for this is [55]. This is the narrowband problem since

$$s(t - \tau) \approx e^{i\omega\tau} s(t).$$

2. Radar Filtering Application

R. Schmidt, in his 1981 Stanford thesis, is responsible for developing the MUSIC method. Actually his results were given at a Rome Air Force Base meeting in 1979. If the true covariance is

$$R = E(zz^*)$$
$$= \sum_{i=1} p_i d(\theta_i) d^*(\theta_i) + I$$
$$= V(\theta) P V(\theta)^* + I,$$

then look at the eigenvalues of R, assuming they are ordered in decreasing size. Note that all eigenvalues are real since R is Hermitian. Then the eigenvalues will behave as depicted in Figure 13.8. The eigenvectors after the rth will be perpendicular to all θs with eigenvalue $= 1$ since

$$d^*(\theta_i) k_i = 0 \quad \Rightarrow Rk = k.$$

Now we know the sample covariance, $\hat{R} = \frac{1}{N} \sum zz^*$, and we can find λ_i and f_i eigenvalues and eigenvectors. Take the r largest eigenvalues, and let

$$\eta = \text{span } ((f_{r+1}) \cdots f_n)$$

and

$$\mathcal{P} : \mathcal{C}^n \to \eta,$$

i.e., $\mathcal{P}$ is a projection. Now, form the following Schmidt statistic:

$$f(\theta) = \frac{1}{d^*(\theta) \mathcal{P}^* \hat{R} \mathcal{P} d(\theta)}.$$

This function "blows up" at the estimated θ locations since $\mathcal{P}$ annihilates $d(\theta)$ there. A typical plot of the $f(\theta)$ function might look like that shown in Figure 13.9. This adds power to the estimation of θ through the

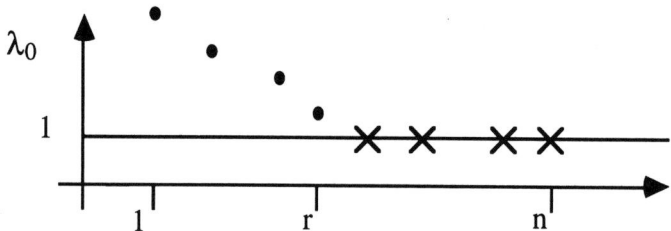

Figure 13.8: Ordered eigenvalues of R.

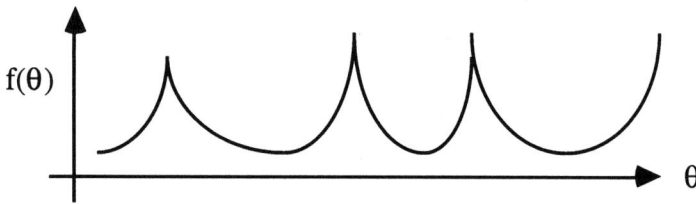

Figure 13.9: Estimation of radar angles using the Schmidt statistic.

number of looks because the number of looks increases the signal-to-noise ratio. This process is called superresolution. Notice that we never had to find the a_is explicitly. This estimator achieves the Rao-Cramer bound for the error variance for high signal-to-noise ratio, even for multiple targets, within a tenth of a beamwidth.

2.1 Rao-Cramer Bound

The Rao-Cramer bound may be described as follows. Suppose there is a random variable $p(x, \theta)$, where x is the index and θ is the mean. If $t(x)$ estimates θ, the following is a heuristic proof. We assume $E(t(x)) = \theta$, unbiasedness.

$$E(t(x)) = \int_{-\infty}^{\infty} xp(x,\theta)dx$$

2. Radar Filtering Application

$$\Rightarrow 0 = \int_{-\infty}^{\infty} (t(x) - \theta) p(x, \theta) dx$$

$$\Rightarrow \int_{-\infty}^{\infty} p(x, \theta) dx = \int_{-\infty}^{\infty} (t(x) - \theta) \frac{1}{p} \frac{dp}{d\theta} p(x, \theta) dx = 1$$

from Schwarz $(Eab)^2 \leq Ea^2 Eb^2$

$$\Rightarrow 1 \leq \int_{-\infty}^{\infty} (t(x) - \theta)^2 p(x, \theta) dx \int_{-\infty}^{\infty} \left(\frac{\partial}{\partial \theta} \ln p \right)^2 p(x, \theta) dx.$$

Hence,

$$\text{var } (t - \theta) \geq \frac{1}{E([\frac{\partial}{\partial \theta} \ln p)^2]},$$

where the term in the denominator of the last relation becomes the observability matrix (or Fisher information matrix) for multidimension problems.

$$E\{(t_i - \theta_i)(t_j - \theta_j)\} \leq W^{-1}$$

$$w_{ij} = E \left\{ \frac{\partial}{\partial \theta_i} \ln p \frac{\partial}{\partial \theta_j} \ln p \right\}.$$

Under some conditions the MLE will attain equality with the Rao-Cramer bound. The reference for such a condition is found in [56].

What happens with random fields (for example, the ocean) where time and space are variables? To find out, see [57] and the random field references in the bibliography of [46]. Wood shows that construction of 2-D Markov fields is not simply a combination of independent Gaussian inputs following Levy's ideas; see [59]. Levy has many important and seminal ideas in Markov process theory in the 1920s through the 1960s.

Lecture 14

Random Fields

Consider a discrete-time Markov process of a random walk in r dimensions, where $r \geq 3$. This process is characterized as follows.

$$\text{Let } \mathcal{F}_s = \min \sigma(x_r \leq s)$$

be the minimum σ field.

$$P(x_t \in B \mid \mathcal{F}_s) = P(x_t \in B \mid x_s)$$
$$= \sum_{y \in B} p(t-s, x, y),$$

where $p(t-s, x, y)$ is the state transition probability

$$p(t-s, x, y) = P_{xy}^{t-s}$$
$$= \Pr(x \text{ goes to } y \text{ in time } t-s).$$

Example 14.1 *Chapman-Kolmogorov equation $p^t = p^s p^{t-s}$ is given by*

$$p(t+s, x, z) = \int p(t, x, y)\, p(s, y, z) dy.$$

For each state, define

$$K(x,y) = \sum_{n=0}^{\infty} P_{xy}^n$$

$$= P_{xy}^0 + P_{xy}^1 + P_{xy}^2 + \cdots$$

$$= \Pr(x \to y \text{ in 0 steps}) + \Pr(x \to y \text{ in 1 step}) +$$

$$\Pr(x \to y \text{ in 2 steps}) + \cdots$$

$$= \sum_{n=0}^{\infty} E(I_{[x_n=y]} \mid x_0 = x),$$

where

$$I_{[x_i=y]} = \begin{cases} 1, & \text{if } x_i = y \\ 0, & \text{otherwise.} \end{cases}$$

$K(x,y)$ therefore counts the expected number of visits to y given x_0 starts at x.

Theorem 14.1 *Chapman-Kolmogorov discrete equation:*

$$p(x_n = y \mid x_0 = x) = \sum_{\forall z} p(x_m = z \mid x_0 = x) p(x_{n-m} = y \mid x_0 = z),$$

for $n > m > 0$.

<u>Proof:</u> Since $\{x_n = y\} = \bigcup_{\forall z}\{x_n = y, x_m = z\}$.
The state transition probability matrix P_{xy}^n can be related to the $K(x,y)$ matrix:

$$P_{xy} K(x,y) = P_{xy}^1 \sum_{n=0}^{\infty} P_{xy}^n$$

$$= \sum_{n=0}^{\infty} P_{xy}^{n+1} = K(x,y) - I,$$

Random Fields

where I is the identity matrix

$$\Rightarrow (I - P)K = I,$$

which is an infinite-dimension matrix equation.

Properties of P_{xy}:
(1) elements of $P_{xy} \geq 0$.
(2) elements of rows of P_{xy} sum to 1.

Example 14.2 *Approximation to Brownian motion. A random step is taken in 3-D space according to the toss of a six-sided die. Then the following outcomes of a roll result in a unit move along one of the coordinate axes: $1 \Rightarrow \vec{e_1} = (1,0,0)$, $2 \Rightarrow \vec{e_2} = (-1,0,0)$, $3 \Rightarrow \vec{e_3} = (0,1,0)$, $4 \Rightarrow \vec{e_4} = (0,-1,0)$, $5 \Rightarrow \vec{e_5} = (0,0,1)$, $6 \Rightarrow \vec{e_6} = (0,0,-1)$. Then the transition probabilities are given by*

$$P_{\vec{r_1},\vec{r_2}} = \frac{1}{6} \sum_{j=1}^{6} \delta_{\vec{r_1},\vec{r_2}+\vec{e_j}},$$

where

$$\delta_{\vec{r_1},\vec{r_2}} = \begin{cases} 1, & \text{if } \vec{r_1} = \vec{r_2} \\ 0, & \text{otherwise.} \end{cases}$$

We would like to construct a Markov process involving triples of integers. First consider a Gaussian process characterized by covariance $R(\vec{k}, \vec{m})$ that is Hermitian and positive semidefinite,

$$\sum_{\vec{k},\vec{m}} \alpha_{\vec{k}} R(\vec{k} - \vec{m}) \bar{\alpha}_{\vec{m}} \geq 0,$$

where stationarity is also assumed.

Example 14.3 *Inverse Fourier transform of $R(\vec{n})$.*

$$R(\vec{n}) = \frac{1}{(2\pi)^3} \int_{-\pi}^{\pi} \int_{-\pi}^{\pi} \int_{-\pi}^{\pi} e^{-i(n_1\theta_1 + n_2\theta_2 + n_3\theta_3)} \mathcal{R}(\theta_1, \theta_2, \theta_3) \, d\theta_1 \, d\theta_2 \, d\theta_3,$$

where $\vec{n} = (n_1, n_2, n_3)$ and $\mathcal{R}(\theta_1, \theta_2, \theta_3)$ is the Fourier transform of $R(\vec{n})$.

Theorem 14.2 (Bockner) $R(\vec{n})$ is positive definite if and only if its Fourier transform is positive, real, and integrable.

Proof:

$$\sum_{\vec{k}} \alpha_{\vec{k}} R(\vec{k}-\vec{m}) \bar{\alpha}_{\vec{m}} = \int \int \int \|\sum e^{i(n_1\theta_1+n_2\theta_2+n_3\theta_3)}\|^2 \mathcal{R} \, d\theta_1 \, d\theta_2 \, d\theta_3.$$

We would like to treat the expected number of visits function, $K(x, y)$, as a covariance in three dimensions. For Example 1.2, the Fourier transform of $(I - P)$ is given by

$$\mathcal{F}(I - P) = 1 - \frac{1}{3}(\cos\theta_1 + \cos\theta_2 + \cos\theta_3).$$

Since $(I - P)K = I$, then

$$\mathcal{F}(K) = \frac{1}{\mathcal{F}(I-P)} = \frac{1}{1 - \frac{1}{3}(\cos\theta_1 + \cos\theta_2 + \cos\theta_3)}$$

is the spectrum of the matrix $K(x, y)$ and is the natural covariance function for a random field:

$$E\{X_{\vec{\alpha}} X_{\vec{\beta}}\} = K(\vec{\alpha} - \vec{\beta}).$$

We would like to show that this random field is Markov.

Example 14.4 Let $\vec{e_1}$, $\vec{e_2}$, $\vec{e_3}$ be unit vectors in 3-D space;

$$X_{\vec{\alpha}} - \frac{1}{3}\sum_{i=1}^{3}(X_{\vec{\alpha}+\vec{e_i}} + X_{\vec{\alpha}-\vec{e_i}}) = U_{\vec{\alpha}}.$$

If X is a Markov process then $U_{\vec{\alpha}}$ cannot be white. We would like to find the spectrum of U such that $S(X) = \mathcal{F}(K)$, where K is the expected number of visits function for Example 1.2. Taking the Fourier transform of the

Random Fields

preceding process:

$$(1 - \tfrac{1}{3}g(\vec{\theta}))\mathcal{X} = \mathcal{U},$$

where $g(\vec{\theta}) = \cos\theta_1 + \cos\theta_2 + \cos\theta_3$. *The power spectra are related by*

$$S(X) = \frac{S(U)}{\|1 - \tfrac{1}{3}g(\vec{\theta})\|^2}.$$

Therefore the required spectrum of the input process is

$$S(U) = 1 - \frac{1}{3}g(\vec{\theta}).$$

The input process $U_{\vec{\alpha}}$ is a correlated process:

$$E\{U_{\vec{\gamma}}, U_{\vec{\alpha}}\} = E\{U_{\vec{\gamma}}, X_{\vec{\alpha}} - U_{\vec{\gamma}} - \tfrac{1}{3}\sum_{i=1}^{3}(X_{\vec{\alpha}+\vec{e}_i} + X_{\vec{\alpha}-\vec{e}_i})\}.$$

From the Markov property, $E\{U_{\vec{\alpha}}, X_{\vec{\beta}}\} = \delta_{\vec{\alpha},\vec{\beta}}$. Therefore

$$E\{U_{\vec{\gamma}}, U_{\vec{\alpha}}\} = \delta_{\vec{\gamma},\vec{\alpha}} - \tfrac{1}{3}\sum_{i=1}^{3}(\delta_{\vec{\gamma},\vec{\alpha}+\vec{e}_i} + \delta_{\vec{\gamma},\vec{\alpha}-\vec{e}_i}),$$

which is not uncorrelated.

The preceding is a Gaussian process indexed by a 3-D lattice.
<u>Question</u>: What is the prediction problem for a random field?
With the covariance as defined, the estimate is

$$\hat{X}_{\vec{\beta}} = E\{X_{\vec{\beta}}y'\}E\{yy'\}^{-1}y.$$

The solution is related to the Dirichlet problem. That is, we want to solve the predictor inside a box when the values on the boundary are known.

<u>Required properties:</u>
(1) initial stochastic process symmetric.
(2) expected number of visits function, $K(x,y)$, is integrable.

Example 14.5 *Consider the process*

$$X_\alpha - \tfrac{1}{2}(X_{\alpha+1} + X_{\alpha-1}) = U_\alpha.$$

By the Markov property, $E\{U_\alpha X_\beta\} = \delta_{\alpha\beta}$. Multiplying by X_γ and taking the expected value:

$$E\{X_\alpha X_\gamma\} - \tfrac{1}{2}(E\{X_{\alpha+1}X_\gamma\} + E\{X_{\alpha-1}X_\gamma\}) = \delta_{\alpha\gamma}.$$

Letting $f(a-b) = E\{X_a X_b\}$, there corresponds an infinite system of equations:

$$\begin{bmatrix} -\tfrac{1}{2} & 1 & -\tfrac{1}{2} & 0 & 0 & \cdots & 0 \\ 0 & -\tfrac{1}{2} & 1 & -\tfrac{1}{2} & 0 & \cdots & 0 \\ 0 & 0 & -\tfrac{1}{2} & 1 & -\tfrac{1}{2} & \cdots & 0 \\ \vdots & \vdots & \vdots & \vdots & \vdots & \vdots & \vdots \end{bmatrix} \begin{bmatrix} f(-1) \\ f(0) \\ f(1) \\ \vdots \end{bmatrix} = \begin{bmatrix} 1 \\ 0 \\ 0 \\ \vdots \end{bmatrix}$$

or $(I - P)$ annihilates f except at the boundary. In this case $K(\alpha) = \tfrac{1}{2\pi}\int \exp(-i\theta\alpha)\tfrac{1}{1-\tfrac{1}{2}\cos\theta}\, d\theta$. The solution is $f(\alpha - \gamma) = \tfrac{K(\alpha-\gamma)}{K(0)}$.

Applications:

1. image processing.

2. weather forecasting.

This means that if we want to find the predictor we have to solve this boundary value problem. Note this will be much more complicated in higher dimensions.

Bibliography

[1] M. Loeve, *Probability Theory*, Vols. 1 and 2, Springer-Verlag, Berlin, 1975.

[2] F.R. Gantmacher, *The Theory of Matrices*, Vol. 1, Chelsea, New York, 1960.

[3] M. Abramowitz and I. Stegun, *Handbook of Mathematical Functions*, Dover, New York, 1965.

[4] R. Bellman, *Introduction to Matrix Analysis*, McGraw-Hill, New York, 1960.

[5] D. Knuth, *The Art Of Computer Programming*, Vol. 2, Addison-Wesley, Reading, Mass., 1969.

[6] K.D. Senne, "Machine Independent Monte Carlo Evaluation of the Performance of Dynamic Stochastic Systems," *Stochastics*, 1(3), 1973, 215–238.

[7] T.W. Anderson, *An Introduction to Multivariate Statistical Analysis*, 2d edition, John Wiley, New York, 1984.

[8] R. Muirhead, *Aspects of Multivariate Statistical Theory*, John Wiley, New York, 1982.

[9] K. Mardia, J. Kent, and J. Bibby, *Multivariate Analysis*, Academic Press, New York, 1979.

[10] N.R. Goodman, "Statistical Analysis Based on Certain Multivariate Complex Gaussian Distributions," *Ann. Math Statistics*, 34, 1963, 152–177.

[11] E. Lehman, *Testing Statistical Hypotheses*, John Wiley, New York, 1959.

[12] H.S. Wall, *Continued Fractions*, Chelsea, New York, 1967.

[13] O. Perron, *Die Lehre von den Kettenbruchen*, Teubner, Stuttgart, 1954.

[14] R.S. Bucy, "A Priori Bounds for the Riccati Equation," *Proc. 6th Berkeley Symposium on Mathematical Statistics and Probability*, vol. 3, Probability Theory, Univ. of California Press, Berkeley, 1970, pp. 645–656.

[15] R.E. Kalman and R.S. Bucy, "New Results in Linear Filtering and Prediction Theory," *J. Basic Eng. ASME*, Series D, 1961, 95–108.

[16] J.W. Follin, Jr. and A.G. Carlton, "Recent Developments in Fixed and Adaptive Filtering," *Proc. Second AGARD Guided Missiles Seminar (Guidance and Control)*, AGARDograph 21, September 1956.

[17] C.C. Carathéodory, *Calculus of Variations and Partial Differential Equations of the First Order*, 2d edition, Chelsea, New York, 1982.

[18] V.I. Arnold, *Mathematical Methods of Classical Mechanics*, translated by K. Vogtmann and A. Weinstein, Springer-Verlag, New York, 1984.

[19] N. Wiener, *Extrapolation, Interpolation and Smoothing of Stationary Time Series*, John Wiley, New York, 1949.

[20] R.E. Kalman, "A New Approach to Linear Filtering and Prediction Theory," *J. Basic Eng., ASME*, 82, 1960, 35–45.

[21] J. Burg, "Maximum Entropy Spectral Analysis," thesis, Stanford University, Stanford, Calif., 1975.

Bibliography

[22] J. Clairbout, *Fundamentals of Geophysical Data Processing*, McGraw-Hill, New York, 1976.

[23] R.S. Bucy, "Identification and Filtering," *Math. Systems Theory*, 16(4), 1983, 307–317.

[24] R.E. Bellman, *Introduction to Matrix Analysis*, McGraw-Hill, New York, 1960.

[25] J.D. Markle and A.H. Gray, Jr., *Linear Prediction of Speech*, Springer-Verlag, New York, 1976.

[26] G. Alengrin, R.S. Bucy, J.M.F. Moura, J. Pagés, and M.I. Ribeiro, "ARMA Identification," *J. Optimization Theory and Applications*, 55(3), 1987, 345–357.

[27] J.P. Burg, D.G. Luenberger, and D.L. Wenger, "Estimation of Structured Covariance Matrices," invited paper, *Proceedings IEEE*, 70(9), 1982, 963–974.

[28] G.C. Newton, L.A. Gould, and J.F. Kaiser, *Analytical Design of Linear Feedback Controls*, John Wiley, New York, 1957.

[29] R.S. Bucy, "Linear and Nonlinear Filtering," invited paper, *Proceedings of the IEEE*, 58 (6), June 1970, 854–864.

[30] R.S. Bucy, "A Priori Bounds for the Riccati Equation," *Proceedings of the 6th Berkeley Symposium on Probability and Statistics*, University of California Press, Berkeley, 1970, pp. 645–656.

[31] W. Hahn, *Theorie und Anwendung der Direkten Methoden von Liapounov*, Springer-Verlag, Berlin, 1959.

[32] J.P. La Salle and S. Lefshetz, *Stability by Liapounov's Direct Method*, Academic Press, New York, 1961.

[33] R.E. Kalman and J. Bertram, "Control Systems Analysis and Design via the 'Second Method' of Lyapounov," *ASME J. Basic Engineering*, Series D, 1960, 371–393.

[34] I.M. Gelfand and A.M. Yaglom, "Calculation of the Amount of Information About One Random Contained in Another," *Trans. Am. Math. Soc.*, 1959.

[35] M.S. Pinsker, *Information and Information Stability of Random Variables and Random Processes*, Holden Day, San Francisco, 1964.

[36] T.E. Duncan, "On the Calculation of the Mutual Information," *SIAM J. Applied Math.*, 19(1), 1970, 215–220.

[37] R.L. Stratonovich, *Information Theory* [in Russian], Soviet Radio, Moscow, 1975.

[38] I.J. Glados and D.E. Gustafson, "Information and Distortion in Filtering Theory," *IEEE Trans. Information Theory*, IT-23(2), 1977, 183–193.

[39] R.S. Bucy, "Information and Filtering," *Information Sciences*, 18, 1979, 179–187.

[40] R.S. Bucy, "Distortion Rate Theory and Filtering," *IEEE Trans. Information Theory*, IT-28, 1982, 336–339.

[41] R.S. Bucy, D. Rappaport, and L.M. Silverman, "Invariant Directions for the Riccati Equation," *IEEE Trans. Auto. Control*, 5, 1970, 535–540.

[42] R.S. Bucy and P.D. Joseph, *Filtering for Stochastic Processes with Applications to Guidance*, Tracts in Applied Mathematics, Interscience, New York, 1968.

[43] A.N. Kolmogorov, "Über die Analytischen Methoden in der Wahrscheinlichkeitsrechnung," *Math. Annalen*, 5, 1931, 3–14.

[44] C. Hecht, thesis, Aerospace Engineering Dept., University of Southern California, Los Angeles, 1972.

Bibliography

[45] R.S. Bucy, K.D. Senne, and C. Hecht, *An Engineer's Guide to Building Nonlinear Filters*, Frank J. Seiler Research Lab. Report, Vol 1,2, AD-746922, USAF, Colorado Springs, 1972.

[46] R.S. Bucy and P.D. Joseph, *Filtering for Stochastic Processes with Applications to Guidance*, 2d edition, Chelsea, New York, 1987.

[47] R.S. Bucy, D.S. Miller, and M.J. Merritt, "Hybrid Computer Synthesis of Optimal Discrete Nonlinear Filters," *Proceedings of the 2d Symposium on Nonlinear Estimation*, La Jolla, Calif., September 13–15, 1973; *Stochastics*, 2, 1973, 151–212.

[48] R.S. Bucy, L. Basanez, P. Brunet, R. Huber, D. Miller, and J. Pages, "A Hybrid Computer Optimal Filter," *Proceedings of the 6th Symposium on Nonlinear Estimation and Applications*, San Diego, Calif., September 1975.

[49] J.W. Goodman, *An Introduction to Fourier Optics*, McGraw-Hill, New York, 1968.

[50] R.S. Bucy, A.J. Mallinckrodt, and H. Youssef, "High Speed Convolution of Periodic Functions," *SIAM J. Math. Analysis*, 8, 1977, 619–625.

[51] R.S. Bucy, W. Steier, C.P. Christensen, A. Sawchuk, and D. Drake, "Feasibility Study of Device Synthesis of Nonlinear Filters," *Technical Report USC-EE 494*, University of Southern California, Los Angeles, 1978.

[52] R.S. Bucy and H. Gautier, "Saw Nonlinear Filter Synthesis," *DASM Report*, Thomson CSF, 84/001/405, July 1984.

[53] R.S. Bucy, "Nonlinear Filtering Algorithms for Vector Processing Machines," *Computers and Mathematics*, 6(3), 1980, 317–338.

[54] S. Perrenod and R.S. Bucy, "Supercomputer Performance for Convolution," *Proceedings of the 1st International Conference on Supercomputers*, St. Petersburg, Fla., December 1985, IEEE Society Press, New York, 1986, 301–309.

[55] G.W. Stimpson, *Airborne Radar*, Hughes Aircraft, El Segundo, Calif., 1983.

[56] H. Cramer, *Random Variables and Probability Distributions*, Cambridge University Press, Cambridge, U.K., 1937.

[57] J.W. Woods, "Two Dimensional Markov Estimation," *IEEE Trans. Information Theory*, IT-22(5), 1976, 50–59.

[58] R.S. Bucy and K.D. Senne, "Digital Realizations of Optimal Discrete Time Nonlinear Filters," *Proceedings of Symposium on Nonlinear Filtering*, September 1970, Western Periodicals, Los Angeles.

[59] P. Levy, *Processus Stochastiques et Mouvement Brownien*, 2d, Gautier-Villars, Paris, 1965.

[60] E.B. Dynkin, "Markov Processes and Random Fields," *Bull. Am. Math. Soc.*, 3(3), November 1980, 975–999.

[61] J.W. Woods, "Two-Dimensional Discrete Markovian Fields," *IEEE Trans. Information Theory*, IT-18(2), March 1972, 232–240.